Linear Algebra for Pattern Processing

Projection, Singular Value Decomposition, and Pseudoinverse

Synthesis Lectures on Signal Processing

Synthesis Lectures in Signal Processing publishes 80- to 150-page books on topics of interest to signal processing engineers and researchers. The Lectures exploit in detail a focused topic. They can be at different levels of exposition—from a basic introductory tutorial to an advanced monograph—depending on the subject and the goals of the author. Over time, the Lectures will provide a comprehensive treatment of signal processing. Because of its format, the Lectures will also provide current coverage of signal processing, and existing Lectures will be updated by authors when justified.
Lectures in Signal Processing are open to all relevant areas in signal processing. They will cover theory and theoretical methods, algorithms, performance analysis, and applications. Some Lectures will provide a new look at a well established area or problem, while others will venture into a brand new topic in signal processing. By careful reviewing the manuscripts we will strive for quality both in the Lectures' contents and exposition.

Linear Algebra for Pattern Processing: Projection, Singular Value Decomposition, and Pseudoinverse
Kenichi Kanatani
2021

Signals and Systems: A One Semester Modular Course
Khalid Sayood
2021

Smartphone-Based Real-Time Digital Signal Processing, Third Edition
Nasser Kehtarnavaz, Abhishek Sehgal, Shane Parris, and Arian Azarang
2020

Anywhere-Anytime Signals and Systems Laboratory: from MATLAB to Smartphones, Third Edition
Nasser Kehtarnavaz, Fatemeh Saki, Adrian Duran, and Arian Azarang
2020

Reconstructive-Free Compressive Vision for Surveillance Applications
Henry Braun, Pavan Turaga, Andreas Spanias, Sameeksha Katoch, Suren Jayasuriya, and Cihan Tepedelenlioglu
2019

Smartphone-Based Real-Time Digital Signal Processing, Second Edition
Nasser Kehtarnavaz, Abhishek Sehgal, Shane Parris
2018

Anywhere-Anytime Signals and Systems Laboratory: from MATLAB to Smartphones, Second Edition
Nasser Kehtarnavaz, Fatemeh Saki, and Adrian Duran
2018

Anywhere-Anytime Signals and Systems Laboratory: from MATLAB to Smartphones
Nasser Kehtarnavaz and Fatemeh Saki
2017

Smartphone-Based Real-Time Digital Signal Processing
Nasser Kehtarnavaz, Shane Parris, and Abhishek Sehgal
2015

An Introduction to Kalman Filtering with MATLAB Examples
Narayan Kovvali, Mahesh Banavar, and Andreas Spanias
2013

Sequential Monte Carlo Methods for Nonlinear Discrete-Time Filtering
Marcelo G.S. Bruno
2013

Processing of Seismic Reflection Data Using MATLAB™
Wail A. Mousa and Abdullatif A. Al-Shuhail
2011

Fixed-Point Signal Processing
Wayne T. Padgett and David V. Anderson
2009

Advanced Radar Detection Schemes Under Mismatched Signal Models
Francesco Bandiera, Danilo Orlando, and Giuseppe Ricci
2009

DSP for MATLAB™ and LabVIEW™ IV: LMS Adaptive Filtering
Forester W. Isen
2009

DSP for MATLAB™ and LabVIEW™ III: Digital Filter Design
Forester W. Isen
2008

DSP for MATLAB™ and LabVIEW™ II: Discrete Frequency Transforms
Forester W. Isen
2008

DSP for MATLAB™ and LabVIEW™ I: Fundamentals of Discrete Signal Processing
Forester W. Isen
2008

The Theory of Linear Prediction
P. P. Vaidyanathan
2007

Nonlinear Source Separation
Luis B. Almeida
2006

Spectral Analysis of Signals: The Missing Data Case
Yanwei Wang, Jian Li, and Petre Stoica
2006

Linear Algebra for Pattern Processing: Projection, Singular Value Decomposition, and Pseudoinverse
Kenichi Kanatani

ISBN: 978-3-031-01416-1 paperback
ISBN: 978-3-031-02544-0 ebook
ISBN: 978-3-031-00337-0 hardcover

DOI 10.1007/978-3-031-02544-0

A Publication in the Springer series
SYNTHESIS LECTURES ON SIGNAL PROCESSING

Lecture #21
Series Editor: José Moura, *Carnegie Mellon University*
Series ISSN
Print 1932-1236 Electronic 1932-1694

Linear Algebra for Pattern Processing

Projection, Singular Value Decomposition, and Pseudoinverse

Kenichi Kanatani
Okayama University

SYNTHESIS LECTURES ON SIGNAL PROCESSING #21

ABSTRACT

Linear algebra is one of the most basic foundations of a wide range of scientific domains, and most textbooks of linear algebra are written by mathematicians. However, this book is specifically intended to students and researchers of pattern information processing, analyzing signals such as images and exploring computer vision and computer graphics applications. The author himself is a researcher of this domain.

Such pattern information processing deals with a large amount of data, which are represented by high-dimensional vectors and matrices. There, the role of linear algebra is not merely numerical computation of large-scale vectors and matrices. In fact, data processing is usually accompanied with "geometric interpretation." For example, we can think of one data set being "orthogonal" to another and define a "distance" between them or invoke geometric relationships such as "projecting" some data onto some space. Such geometric concepts not only help us mentally visualize abstract high-dimensional spaces in intuitive terms but also lead us to find what kind of processing is appropriate for what kind of goals.

First, we take up the concept of "projection" of linear spaces and describe "spectral decomposition," "singular value decomposition," and "pseudoinverse" in terms of projection. As their applications, we discuss least-squares solutions of simultaneous linear equations and covariance matrices of probability distributions of vector random variables that are not necessarily positive definite. We also discuss fitting subspaces to point data and factorizing matrices in high dimensions in relation to motion image analysis. Finally, we introduce a computer vision application of reconstructing the 3D location of a point from three camera views to illustrate the role of linear algebra in dealing with data with noise. This book is expected to help students and researchers of pattern information processing deepen the geometric understanding of linear algebra.

KEYWORDS

linear spaces, eigenvalues, spectral decomposition, singular value decomposition, pseudoinverse, least-squares solution, Karhunen–Loève expansion, principal component analysis, trifocal tensors

Contents

Preface

Linear algebra is one of the most basic foundations of a wide range of scientific domains, including physics, chemistry, mechanical and electronic engineering, agriculture, economics, and medicine, and is taught in each domain as a first course in almost all universities. Reflecting this fact, most textbooks of linear algebra are intended to such a wide range of students and mostly written by mathematicians. In contrast, this book is specifically intended to students and researchers of pattern information processing, analyzing signals such as images and exploring computer vision and computer graphics applications. The author himself is a researcher of this domain.

Such pattern information processing deals with a large amount of data, which are represented by high-dimensional vectors and matrices. There, the role of linear algebra is not merely numerical computation of large-scale vectors and matrices. In fact, data processing is usually accompanied with "geometric interpretation." For example, we can think of one data set being "orthogonal" to another and define a "distance" between them or invoke geometric relationships such as "projecting" some data onto some space. Such geometric concepts not only help us mentally visualize abstract high-dimensional spaces in intuitive terms but also lead us to find what kind of processing is appropriate for what kind of goals.

With such applications in mind, we provide explanations to those themes of linear algebra that play fundamental roles in patter information processing. First, we take up the concept of "projection" of linear spaces and describe "spectral decomposition," "singular value decomposition," and "pseudoinverse" in terms of projection. As their applications, we discuss least-squares solutions of simultaneous linear equations and covariance matrices of probability distributions of vector random variables that are not necessarily positive definite. We also discuss fitting subspaces to point data and factorizing matrices in high dimensions in relation to motion image analysis. Finally, we introduce a computer vision application of reconstructing the 3D location of a point from three camera views to illustrate the role of linear algebra in dealing with data with noise.

The starting point of all these is the concept of "projection." This is because it implies the concepts of "orthogonality" and "minimum distance," on which pattern information processing is founded. This distinguishes this book from most textbooks of linear algebra, which usually focus on matrix operation and numerical computation. Thus, this book is expected to help students and researchers of pattern information processing deepen the geometric understanding of linear algebra.

This book is assuming that the readers have already studied basics of linear algebra taught in first college courses, such as manipulation of vectors, matrices, and determinants as well as computation of eigenvalues, eigenvectors, and canonical forms of quadratic forms. To help those who feel unsure about them, a "Glossary and Summary" section is given in each chapter, providing brief explanations of mathematical terminologies that appear in that chapter and related topics and applications. Those who may feel the explanations too elementary can find more advanced discussions in "Supplemental Notes" at the end of each chapter.

At the end of the volume, an appendix entitled "Fundamentals of Linear Algebra" is given to summarize rudimentary knowledge of linear algebra and related mathematical facts. Moreover, many basic facts of linear algebra are collected as "Problems" at the end of each chapter and their "Answers" are given at the end of the volume. With these supplements, this book can also be used for refreshing basic knowledge of linear algebra.

This book is expected to be widely read by both general readers who are interested in mathematics and those researchers who are studying and developing pattern information processing systems. This book is also suitable as advanced textbooks of university courses related to pattern information processing.

The author thanks Xavier Pennec of INRIA Sophia Antipolis for helpful comments. Parts of this work are based with permission on the author's work *Seminar on Linear Algebra: Projection, Singular Value Decomposition, Pseudoinverse* (in Japanese), ©Kyoritsu Shuppan Co., Ltd., 2018.

Kenichi Kanatani
April 2021

CHAPTER 1
Introduction

In this book, we introduce basic mathematical concepts of linear algebra that underlie pattern information processing in high dimensions and discuss some applications to 3D analysis of multiple images. The organization of this book is as follows.

1.1 LINEAR SPACE AND PROJECTION

In Chapter 2, we introduce the concepts of *projection* and *rejection* and express them in the form of the *projection matrix*. Projection plays a central role in this book, because it implies both *orthogonality* and *shortest distance*. The themes of the subsequent chapters are all based on these two aspects of projection. First, we define *subspaces*, *orthogonal complements*, and *direct sum decomposition* and then derive concrete expressions of the projection matrix. As an illustration, we show examples of projection onto lines and planes and introduce the *Schmidt orthogonalization* for producing an orthonormal system of vectors by successive projection operations.

1.2 EIGENVALUES AND SPECTRAL DECOMPOSITION

In Chapter 3, we show that a symmetric matrix can be expressed in terms of its *eigenvalues* and *eigenvectors*. The expression is called the *spectral decomposition*. It allows us to convert a symmetric matrix into a diagonal matrix by multiplying it by an *orthogonal matrix* from left and right. This process is called *diagonalization* of a symmetric matrix. We can also express the inverse and powers of a symmetric matrix in terms of its spectral decomposition.

1.3 SINGULAR VALUES AND SINGULAR VALUE DECOMPOSITION

The spectral decomposition is defined only for symmetric matrices, hence only for square matrices. In Chapter 4, we extend it to arbitrary rectangular matrices; we define the *singular value decomposition*, which expresses any matrix in terms of its *singular values* and *singular vectors*. The singular vectors form a basis of the subspace spanned by the columns or the rows, defining a projection matrix onto it.

1.4 PSEUDOINVERSE

A square matrix has its inverse if it is nonsingular. In Chapter 5, we define the *pseudoinverse*, which extends the inverse to an arbitrary rectangular matrix. While the usual inverse is defined in such a way that its product with the original matrix equals the identity, the product of the pseudoinverse with the original matrix is the projection matrix onto the subspace spanned by its columns and rows. Since all the columns and rows of a nonsingular matrix are linearly independent, they span the entire space, and the projection matrix onto it is the identity. In this sense, the pseudoinverse is a natural extension of the usual inverse. Next, we show that vectors, i.e., $n \times 1$ or $1 \times n$ matrices, also have their pseudoinverses. We then point out that we need a special care for computing the pseudoinverse of a matrix whose elements are obtained by measurement in the presence of noise. We also point out that the error in such matrices is evaluated in the *matrix norm*.

1.5 LEAST-SQUARES SOLUTION OF LINEAR EQUATIONS

The pseudoinverse is closely related to the least-squares method for linear equations. In fact, the theory of pseudoinverse has been studied in relation to minimization of the sum of squares of linear equations. The least-squares method usually requires solving an equation, called the *normal equation*, obtained by letting the derivative of the sum of squares be zero. In Chapter 6, we show how a general solution is obtained without using differentiation or normal equations. As illustrative examples, we show the case of multiple equations of one variable and the case of a single multivariate equation.

1.6 PROBABILITY DISTRIBUTION OF VECTORS

In Chapter 7, we regard measurement data that contain noise not as definitive values but as *random variables* specified by probability distributions. The principal parameters that characterize a probability distribution are the *mean* (average) and the *covariance matrix*. In particular, the *normal* (or *Gaussian*) *distribution* is characterized by the mean and the covariance matrix alone. We show that if the probability is not distributed over the entire space but is restricted to some domain, e.g., constrained to be on a planar surface or on a sphere, the covariance matrix becomes singular. In such a case, the probability distribution is characterized by the pseudoinverse of the covariance matrix. We illustrate how this leads to a practical method for comparing computational accuracy of such data.

1.7 FITTING SPACES

In Chapter 8, generalizing line fitting in 2D and plane fitting in 3D to general dimensions, we consider how to fit subspaces and affine spaces to a given set of points in nD. A subspace is a space spanned by vectors starting from the origin, and an *affine space* is a translation of a subspace to a general position. The fitting is done hierarchically: we first fit a lower dimensional space, starting from a 0D space, then determine a space with an additional dimension so that the discrepancy is minimized, and continue this process. This corresponds to what is known as *Karhunen–Loéve expansion* in signal processing and pattern recognition and as *principal component analysis* in statistics. The fitted space is computed from the spectral decomposition of a matrix, which we call the *covariance matrix*, but also can be obtained from its singular value decomposition. We point out that the use of the singular value decomposition is more efficient with smaller computational complexity.

1.8 MATRIX FACTORIZATION

In Chapter 9, we consider the problem of *factorization* of a matrix, i.e., expressing a given matrix $\boldsymbol{A}$ as the product $\boldsymbol{A} = \boldsymbol{A}_1\boldsymbol{A}_2$ of two matrices $\boldsymbol{A}_1$ and, $\boldsymbol{A}_2$. We discuss its relationship to the matrix rank and the singular value decomposition. As a typical application, we describe a technique, called the *factorization method*, for reconstructing the 3D structure of the scene from images captured by multiple cameras.

1.9 TRIANGULATION FROM THREE VIEWS

In Chapter 10, we consider the problem of reconstructing 3D points from three images. The basic principle is to optimally correct observed image points in the presence of noise, in such a way that their rays, or their lines of sight, intersect at a single point in the scene. We show that the three rays intersect in the scene if and only if the *trilinear constraint*, specified by the *trifocal tensor*, is satisfied. We discuss its geometric meaning and point out the role of pseudoinverse in the course of iterative optimal computation.

1.10 FUNDAMENTALS OF LINEAR ALGEBRA

In Appendix A, we summarize fundamentals of linear algebra and related mathematical facts relevant to the discussions in this book. First, we give a formal proof that a linear mapping is written as the product with some matrix. Then, we describe basic facts about the inner product and the norm of vectors. Next, we list formulas that frequently appear in relation to *linear forms*, *quadratic forms*, and *bilinear forms*. Then, we briefly discuss expansion of vectors with respect to an orthonormal basis and least-squares approximation. We also introduce *Lagrange's method of indeterminate multipliers* for computing the maximum/minimum of a function of vectors subject to constraints. Finally, we summarize basics of eigenvalues and eigenvectors and their applications to maximization/minimization of a quadratic form.

CHAPTER 2

Linear Space and Projection

In this chapter, we introduce the concepts of "projection" and "rejection" and express them in the form of the "projection matrix." They play a central role in this book, because it implies both "orthogonality" and "shortest distance." The themes of the subsequent chapters are all based on these two aspects of projection. First, we define "subspaces," "orthogonal complements," and "direct sum decomposition" and then derive concrete expressions of the projection matrix. As an illustration, we show examples of projection onto lines and planes and introduce the "Schmidt orthogonalization" for producing an orthonormal system of vectors using projection matrices.

2.1 EXPRESSION OF LINEAR MAPPING

A linear mapping from the nD space $\mathcal{R}^n$, called the *domain*, to the mD space $\mathcal{R}^m$, is represented by an $m \times n$ matrix $\boldsymbol{A}$ ($\hookrightarrow$ Appendix A.1). One of the basic ways to specify it is to define an *orthonormal basis* $\{\boldsymbol{u}_1, \dots, \boldsymbol{u}_n\}$, i.e., mutually orthogonal unit vectors, in $\mathcal{R}^n$, and their *images*, $\boldsymbol{a}_1, \dots, \boldsymbol{a}_n$, i.e., the mD vectors to which the basis vectors are to be mapped (Fig. 2.1). Then, we see that the matrix $\boldsymbol{A}$ is written as ($\hookrightarrow$ Problem 2.1)

$$\boldsymbol{A} = \boldsymbol{a}_1\boldsymbol{u}_1^\top + \cdots + \boldsymbol{a}_n\boldsymbol{u}_n^\top. \tag{2.1}$$

In fact, if we multiply Eq. (2.1) by $\boldsymbol{u}_i$ from right, we obtain $\boldsymbol{A}\boldsymbol{u}_i = \boldsymbol{a}_i$ due to the orthonormality

$$\boldsymbol{u}_i^\top\boldsymbol{u}_j = \delta_{ij}, \tag{2.2}$$

where δ_{ij} is the *Kronecker delta*, taking the value 1 for $j = i$ and 0 otherwise.

If we use the *natural basis* (also called the *standard* or *canonical basis*) $\{\boldsymbol{e}_1, \dots \boldsymbol{e}_n\}$ for $\{\boldsymbol{u}_1, \dots \boldsymbol{u}_n\}$, where $\boldsymbol{e}_i$ is the nD vector whose ith component is 1 and whose other components are all 0, and if we write the vetor $\boldsymbol{a}_i$ as $\boldsymbol{a}_i = (a_{1i}, \dots, a_{mi})^\top$, Eq. (2.1) has the form

$$\begin{aligned}\boldsymbol{A} &= \begin{pmatrix} a_{11} \\ \vdots \\ a_{m1} \end{pmatrix}\begin{pmatrix} 1 & 0 & \cdots & 0 \end{pmatrix} + \cdots + \begin{pmatrix} a_{1n} \\ \vdots \\ a_{mn} \end{pmatrix}\begin{pmatrix} 0 & \cdots & 0 & 1 \end{pmatrix} \\ &= \begin{pmatrix} a_{11} & \cdots & a_{1n} \\ \vdots & \ddots & \vdots \\ a_{m1} & \cdots & a_{mn} \end{pmatrix}. \end{aligned} \tag{2.3}$$

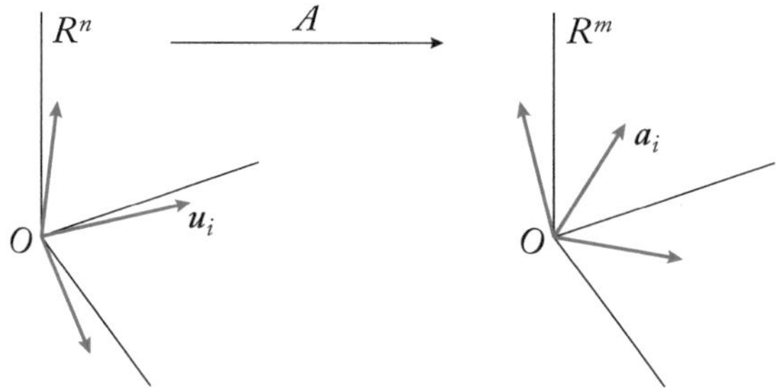

Figure 2.1: The linear mapping that maps an orthonormal basis $\{\boldsymbol{u}_i\}$, $i = 1, \ldots, n$, of $\mathcal{R}^n$ to vectors $\boldsymbol{a}_i$, $i = 1, \ldots, n$, of $\mathcal{R}^m$ is given by the $m \times n$ matrix $\boldsymbol{A} = \sum_{i=1}^{n} \boldsymbol{a}_i \boldsymbol{u}_i^\top$.

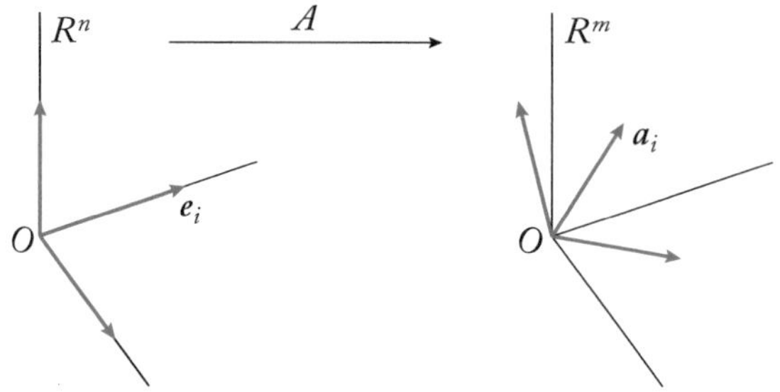

Figure 2.2: The linear mapping that maps the natural basis $\{\boldsymbol{e}_i\}$, $i = 1, \ldots, n$, of $\mathcal{R}^n$ to vectors $\boldsymbol{a}_i = (a_{1i}, \ldots, a_{mi})^\top$, $i = 1, \ldots, n$, of $\mathcal{R}^m$ is given by the $m \times n$ matrix $\boldsymbol{A} = \left(a_{ij}\right)$.

In other words, *the matrix* $\boldsymbol{A}$ *consists of the images* $\boldsymbol{a}_1, \ldots, \boldsymbol{a}_n$ *as its columns in that order* (Fig. 2.2), i.e., $\boldsymbol{A} = \left(a_{ij}\right)$ (shorthand for a matrix whose (i, j) element is a_{ij}).

EXAMPLE: 2D ROTATION

Rotation by angle θ (anti-clockwise) in two dimensions is a linear mapping. The natural basis vectors $\boldsymbol{e}_1 = (1, 0)^\top$ and $\boldsymbol{e}_2 = (0, 1)^\top$ are mapped to $\boldsymbol{a}_1 = (\cos\theta, \sin\theta)^\top$ and $\boldsymbol{a}_2 = (-\sin\theta, \cos\theta)^\top$, respectively, (Fig. 2.3). Hence, rotation by angle θ is represented by the matrix $\boldsymbol{R}(\theta) = \begin{pmatrix} \cos\theta & -\sin\theta \\ \sin\theta & \cos\theta \end{pmatrix}$.

2.2 SUBSPACES, PROJECTION, AND REJECTION

Let $\boldsymbol{u}_1, \ldots, \boldsymbol{u}_r$ be a set of r linearly independent vectors in $\mathcal{R}^n$ (see the grossary in Section 2.6). The set $\mathcal{U} \subset \mathcal{R}^n$ of all linear combinations of these vectors is called the *subspace* of dimension r *spanned by* $\boldsymbol{u}_1, \ldots, \boldsymbol{u}_r$. In particular, the subspace spanned by one vector is a line that extends along it, and the subspace spanned by two vectors is the plane that passes through them.

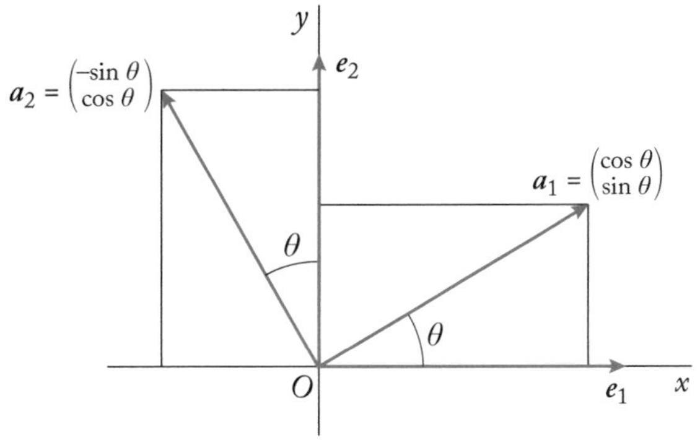

Figure 2.3: The natural basis vectors $\boldsymbol{e}_1 = (1,0)^\top$ and $\boldsymbol{e}_2 = (0,1)^\top$ are mapped to $\boldsymbol{a}_1 = (\cos\theta, \sin\theta)^\top$ and $\boldsymbol{a}_2 = (-\sin\theta, \cos\theta)^\top$, respectively, after a rotation by angle θ.

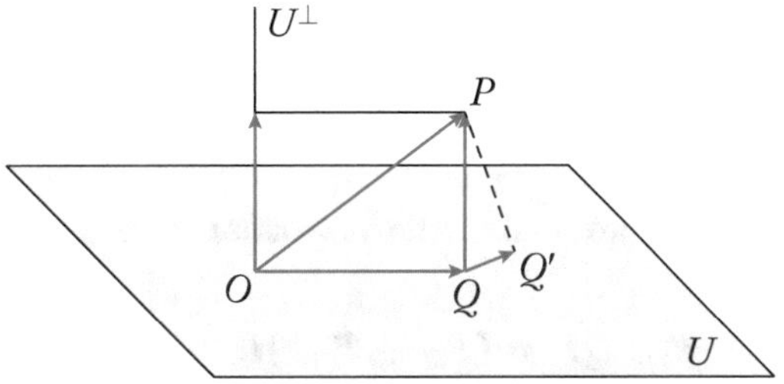

Figure 2.4: The projection Q of point P onto the subspace $\mathcal{U}$ is the point of $\mathcal{U}$ closest to P. The vector $\overrightarrow{QP} \in \mathcal{U}^\perp$ is the rejection from $\mathcal{U}$.

Given a point P in $\mathcal{R}^n$ and a subspace $\mathcal{U} \subset \mathcal{R}^n$, the point $Q \in \mathcal{U}$ defined so that $\overrightarrow{PQ}$ is orthogonal to $\mathcal{U}$ is called the *projection* of P onto $\mathcal{U}$ (Fig. 2.4), and $\overrightarrow{QP}$ is said to be the *rejection* of Q from $\mathcal{U}$ (see the grossary in Section 2.6). If we move the point Q to another point Q' of $\mathcal{U}$, we see from the Pythagorean theorem ($\hookrightarrow$ Eq. (A.12) in Appendix A.2) that,

$$\|PQ'\|^2 = \|PQ\|^2 + \|QQ'\|^2 > \|PQ\|^2. \tag{2.4}$$

In other words, *the projection Q is the closest point of $\mathcal{U}$ from point P* ($\hookrightarrow$ Problem 2.2).

These facts are summarized as

$$\overrightarrow{OP} = \overrightarrow{OQ} + \overrightarrow{QP}, \qquad \overrightarrow{OQ} \in \mathcal{U}, \qquad \overrightarrow{QP} \in \mathcal{U}^\perp, \tag{2.5}$$

where $\mathcal{U}^\perp$ is the set of all vectors orthogonal to $\mathcal{U}$, called the *orthogonal complement* of $\mathcal{U}$, which is also a subspace of $\mathcal{R}^n$. Thus, any vector of $\mathcal{R}^n$ is expressed as the sum of its projection onto $\mathcal{U}$ and the rejection from it. Such an expression is unique and called the *direct sum decomposition* of $\overrightarrow{OP}$ to $\mathcal{U}$ and $\mathcal{U}^\perp$.

2.3 PROJECTION MATRICES

Let $\boldsymbol{P}_{\mathcal{U}}$ be the projection mapping onto subspace $\mathcal{U}$, and $\boldsymbol{P}_{\mathcal{U}^\perp}$ the projection mapping onto its orthogonal complement $\mathcal{U}^\perp$. By definition,

$$\boldsymbol{P}_{\mathcal{U}}\boldsymbol{x} = \begin{cases} \boldsymbol{x} & \boldsymbol{x} \in \mathcal{U} \\ \boldsymbol{0} & \boldsymbol{x} \in \mathcal{U}^\perp \end{cases}, \tag{2.6}$$

$$\boldsymbol{P}_{\mathcal{U}^\perp}\boldsymbol{x} = \begin{cases} \boldsymbol{0} & \boldsymbol{x} \in \mathcal{U} \\ \boldsymbol{x} & \boldsymbol{x} \in \mathcal{U}^\perp \end{cases}. \tag{2.7}$$

If we define an orthonormal basis $\{\boldsymbol{u}_1, \ldots, \boldsymbol{u}_r\}$ of the subspace $\mathcal{U}$, it can be extended to an orthonormal basis $\{\boldsymbol{u}_1, \ldots, \boldsymbol{u}_r, \boldsymbol{u}_{r+1}, \ldots, \boldsymbol{u}_n\}$ of $\mathcal{R}^n$. Equation (2.6) states that $\boldsymbol{P}_{\mathcal{U}}$ maps the orthonormal basis vectors $\{\boldsymbol{u}_1, \ldots, \boldsymbol{u}_n\}$ of $\mathcal{R}^n$ to $\boldsymbol{u}_1, \ldots, \boldsymbol{u}_r, \boldsymbol{0}, \ldots, \boldsymbol{0}$, respectively. Similarly, Eq. (2.7) states that $\boldsymbol{P}_{\mathcal{U}^\perp}$ maps $\{\boldsymbol{u}_1, \ldots, \boldsymbol{u}_n\}$ to $\boldsymbol{0}, \ldots, \boldsymbol{0}, \boldsymbol{u}_{r+1}, \ldots, \boldsymbol{u}_n$, respectively. Hence, from Eq. (2.1), the mappings $\boldsymbol{P}_{\mathcal{U}}$ and $\boldsymbol{P}_{\mathcal{U}^\perp}$ are expressed as matrices

$$\boldsymbol{P}_{\mathcal{U}} = \boldsymbol{u}_1\boldsymbol{u}_1^\top + \cdots + \boldsymbol{u}_r\boldsymbol{u}_r^\top, \tag{2.8}$$

$$\boldsymbol{P}_{\mathcal{U}^\perp} = \boldsymbol{u}_{r+1}\boldsymbol{u}_{r+1}^\top + \cdots + \boldsymbol{u}_n\boldsymbol{u}_n^\top, \tag{2.9}$$

respectively, where $\boldsymbol{P}_{\mathcal{U}}$ and $\boldsymbol{P}_{\mathcal{U}^\perp}$ are called the *projection matrices* onto subspaces $\mathcal{U}$ and $\mathcal{U}^\perp$, respectively.

Since $\overrightarrow{QP} = \boldsymbol{P}_{\mathcal{U}}\overrightarrow{QP} + \boldsymbol{P}_{\mathcal{U}^\perp}\overrightarrow{QP} = (\boldsymbol{P}_{\mathcal{U}} + \boldsymbol{P}_{\mathcal{U}^\perp})\overrightarrow{QP}$ for every point P, we see that

$$\boldsymbol{P}_{\mathcal{U}} + \boldsymbol{P}_{\mathcal{U}^\perp} = \boldsymbol{I}. \tag{2.10}$$

Hence, the identity matrix $\boldsymbol{I}$ is decomposed into the sum of the projection matrix onto the subspace $\mathcal{U}$ and the projection matrix onto its orthogonal complement $\mathcal{U}^\perp$ in the form

$$\boldsymbol{I} = \underbrace{\boldsymbol{u}_1\boldsymbol{u}_1^\top + \cdots + \boldsymbol{u}_r\boldsymbol{u}_r^\top}_{\boldsymbol{P}_{\mathcal{U}}} + \underbrace{\boldsymbol{u}_{r+1}\boldsymbol{u}_{r+1}^\top + \cdots + \boldsymbol{u}_n\boldsymbol{u}_n^\top}_{\boldsymbol{P}_{\mathcal{U}^\perp}}. \tag{2.11}$$

Note that the identity matrix itself is the projection matrix onto the entire space $\mathcal{R}^n$.

Since the vector $\overrightarrow{OQ} = \boldsymbol{P}_{\mathcal{U}}\overrightarrow{OP}$ on the right side of Eq. (2.5) and the vector $\overrightarrow{QP} = \boldsymbol{P}_{\mathcal{U}^\perp}\overrightarrow{OP}$ are orthogonal to each other, we have $\|\overrightarrow{OP}\|^2 = \|\overrightarrow{OQ}\|^2 + \|\overrightarrow{QP}\|^2$. Hence, we see that

$$\|\boldsymbol{x}\|^2 = \|\boldsymbol{P}_{\mathcal{U}}\boldsymbol{x}\|^2 + \|\boldsymbol{P}_{\mathcal{U}^\perp}\boldsymbol{x}\|^2 \tag{2.12}$$

for an arbitrary vector $\boldsymbol{x}$ (Fig. 2.5).

For the projection matrix $\boldsymbol{P}_{\mathcal{U}}$, the following hold ($\hookrightarrow$ Problem 2.3):

$$\boldsymbol{P}_{\mathcal{U}}^\top = \boldsymbol{P}_{\mathcal{U}}, \tag{2.13}$$

$$\boldsymbol{P}_{\mathcal{U}}^2 = \boldsymbol{P}_{\mathcal{U}}. \tag{2.14}$$

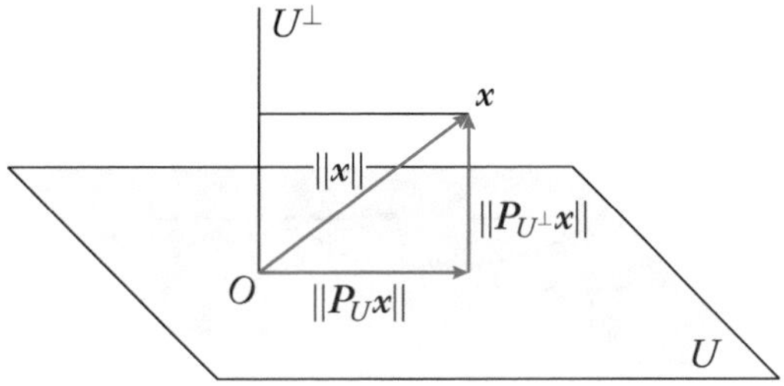

Figure 2.5: The equality $\|x\|^2 = \|P_{\mathcal{U}}x\|^2 + \|P_{\mathcal{U}^\perp}x\|^2$ holds for any vector x.

Equation (2.13) states that $P_{\mathcal{U}}$ is a symmetric matrix, as is evident from the definition of Eq. (2.8). Equation (2.14) states that the projected point is unchanged if it is projected again, which is evident from the meaning of projection. A matrix for which Eq. (2.14) holds is said to be *idempotent*. It can be easily shown that a matrix that is symmetric and idempotent represents the projection matrix onto some subspace (↪ Problem 2.4).

2.4 PROJECTION ONTO LINES AND PLANES

A line l starting from the origin O and extending in the direction of unit vector u is a 1D subspace. The projection matrix P_l onto the line l is given by

$$P_l = uu^\top. \tag{2.15}$$

Hence, the projection of $\overrightarrow{OP}$ onto l is given by

$$uu^\top \overrightarrow{OP} = \langle \overrightarrow{OP}, u \rangle u, \tag{2.16}$$

where and hereafter we denote the inner product of vectors a and b by $\langle a, b \rangle$ $(= a^\top b)$ (↪ Appendix A.2). The right side of Eq. (2.16) is the vector lying on the line l with length $\langle \overrightarrow{OP}, u \rangle$ (Fig. 2.6), signed so that it is positive in the direction of u and negative in the opposite direction. This signed length is called the *projected length*. Thus, we conclude that *the inner product with a unit vector equals the projected length onto the line in that direction.*

A plane Π passing through the origin O having a unit vector n as its surface normal is a subspace of dimension $n-1$. The line along the surface normal n is the orthogonal complement of the plane Π (Strictly speaking, this is a "hyperplane," but we call it simply a "plane" if confusion does not occur). Hence, if P_n is the projection matrix onto Π, Eqs. (2.10) and (2.11) imply

$$P_n = I - nn^\top. \tag{2.17}$$

Thus, the projection of $\overrightarrow{OP}$ onto Π (Fig. 2.7) is given by

$$P_n \overrightarrow{OP} = \overrightarrow{OP} - \langle \overrightarrow{OP}, n \rangle n. \tag{2.18}$$

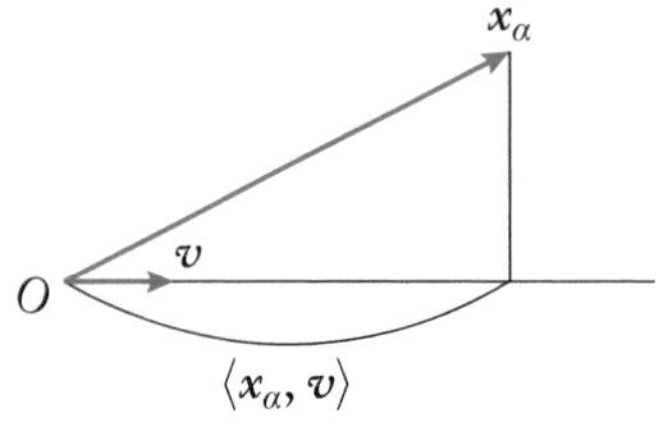

Figure 2.6: The projected length of vector $\overrightarrow{OP}$ onto a line passing through the origin O and extending in the direction of the unit vector $\boldsymbol{u}$ is given by $\langle \overrightarrow{OP}, \boldsymbol{u}\rangle$.

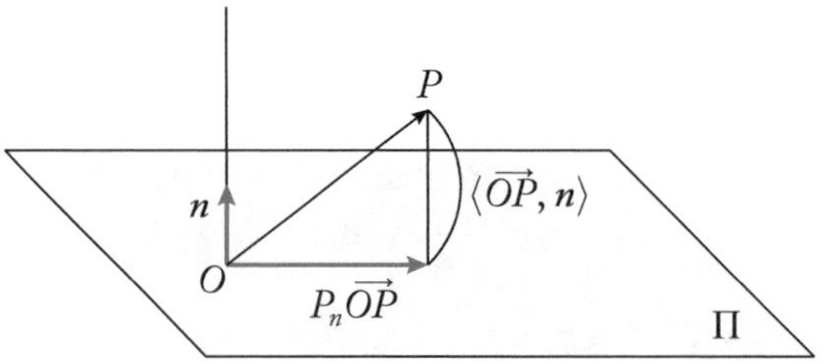

Figure 2.7: Projection of vector $\overrightarrow{OP}$ onto plane Π passing through the origin O and having unit surface normal $\boldsymbol{n}$.

2.5 SCHMIDT ORTHOGONALIZATION

A set of mutually orthogonal unit vectors is said to be an *orthonormal system*. We can convert n given linear independent vectors $\boldsymbol{a}_1, \dots, \boldsymbol{a}_n$ to an orthonormal system $\boldsymbol{u}_1, \dots, \boldsymbol{u}_n$ as follows. First, let $\boldsymbol{u}_1 = \boldsymbol{a}_1/\|\boldsymbol{a}_1\|$. From Eq. (2.17), the projection matrix onto the subspace orthogonal to $\boldsymbol{u}_1$, i.e., its orthogonal complement, is $\boldsymbol{I} - \boldsymbol{u}_1\boldsymbol{u}_1^\top$. We project $\boldsymbol{a}_2$ onto it to extract the component orthogonal to $\boldsymbol{u}_1$. We obtain

$$\boldsymbol{a}_2' = (\boldsymbol{I} - \boldsymbol{u}_1\boldsymbol{u}_1^\top)\boldsymbol{a}_2 = \boldsymbol{a}_2 - \langle \boldsymbol{u}_1, \boldsymbol{a}_2\rangle \boldsymbol{u}_1. \tag{2.19}$$

Its normalization $\boldsymbol{u}_2 = \boldsymbol{a}_2'/\|\boldsymbol{a}_2'\|$ is a unit vector orthogonal to $\boldsymbol{u}_1$. By the same argument, the projection matrix onto the subspace orthogonal to $\boldsymbol{u}_1$ and $\boldsymbol{u}_2$, i.e., its orthogonal complement, is $\boldsymbol{I} - \boldsymbol{u}_1\boldsymbol{u}_1^\top - \boldsymbol{u}_2\boldsymbol{u}_2^\top$. We project $\boldsymbol{a}_3$ onto it to obtain

$$\boldsymbol{a}_3' = (\boldsymbol{I} - \boldsymbol{u}_1\boldsymbol{u}_1^\top - \boldsymbol{u}_2\boldsymbol{u}_2^\top)\boldsymbol{a}_3 = \boldsymbol{a}_3 - \langle \boldsymbol{u}_1, \boldsymbol{a}_3\rangle \boldsymbol{u}_1 - \langle \boldsymbol{u}_2, \boldsymbol{a}_3\rangle \boldsymbol{u}_2, \tag{2.20}$$

which is orthogonal to both $\boldsymbol{u}_1$ and $\boldsymbol{u}_2$; its normalization $\boldsymbol{u}_3 = \boldsymbol{a}_3'/\|\boldsymbol{a}_3'\|$ is a unit vector orthogonal to both $\boldsymbol{u}_1$ and $\boldsymbol{u}_2$. Repeating the same argument, we see that if we already have mutually orthogonal unit vectors $\boldsymbol{u}_1, \dots, \boldsymbol{u}_{k-1}$, the projection matrix onto the subspace orthogonal to $\boldsymbol{u}_1, \dots, \boldsymbol{u}_{k-1}$, i.e., its orthogonal complement, is $\boldsymbol{I} - \boldsymbol{u}_1\boldsymbol{u}_1^\top - \cdots - \boldsymbol{u}_{k-1}\boldsymbol{u}_{k-1}^\top$. We project $\boldsymbol{a}_k$

onto it to obtain

$$a'_k = (I - u_1 u_1^\top - \cdots - u_{k-1} u_{k-1}^\top) a_k = a_k - \langle u_1, a_k \rangle u_1 - \cdots - \langle u_{k-1}, a_k \rangle u_k, \tag{2.21}$$

which is orthogonal to all of $u_1, \ldots, u_{k-1}$; its normalization $u_k = a'_k / \|a'_k\|$ is a unit vector orthogonal to all of $u_1, \ldots, u_{k-1}$. Iterating this process for $k = 1, \ldots, n$, we end up with an orthonormal system $u_1, \ldots, u_n$. This procedure is called the (*Gram–*)*Schmidt orthogonalization*.

2.6 GLOSSARY AND SUMMARY

basis: A set of linearly independent vectors $u_1, \ldots, u_n$ such that any element of the linear space is expressed as their linear combination. The number n is called the "dimension" of the linear space.

direct sum decomposition: Expressing every element of $\mathcal{R}^n$ as the sum of an element of a subspace $\mathcal{U}$ and its orthogonal complement $\mathcal{U}^\perp$. The expression is unique.

idempotent matrix: A matrix whose any power is equal to it is "idempotent." The projection matrix $P_\mathcal{U}$ is idempotent.

Kronecker delta: The symbol δ_{ij} that takes value 1 for $i = j$ and value 0 for $i \neq j$.

linear combination: The expression of the sum of scalar multiples in the form of $c_1 u_1 + \cdots + c_m u_m$. The set of all linear combinations of vectors $u_1, \ldots, u_m$ is called the linear space "spanned" by them.

linear independece: Vectors $u_1, \ldots, u_m$ are "linearly independent" if $c_1 u_1 + \cdots + c_m u_m = 0$ does not hold unless $c_1 = \cdots = c_r = 0$. Otherwise, they are "linearly dependent."

linear mapping: A mapping between linear spaces such that the sum corresponds to the sum and the scalar multiple to the scalar multiple in the form $f(u + v) = f(u) + f(v)$ and $f(cu) = cf(u)$. The space for which the mapping is defined is called its "domain," and the mapped element is called its "image." In the nD space $\mathcal{R}^n$, the image has the form of the product of a vector u and a matrix A.

linear space: A set of elements for which sums and scalar multiples are defined, also called "vector space." Its elements are called "vectors."

natural basis: A set of nD vectors e_i, $i = 1, \ldots, n$, such that the ith component of e_i is 1 and other components are all 0. It is also known as the "standard basis" or the "canonical basis." It is a typical orthonormal basis.

orthogonal complement: For a subspace $\mathcal{U}$ of $\mathcal{R}^n$, the subspace consisting of vectors orthogonal to all $u \in \mathcal{U}$ is called its "orthogonal complement" and is denoted by $\mathcal{U}^\perp$.

orthonormal basis: A basis consisting of mutually orthogonal unit vectors.

orthonormal system: A set of mutually orthogonal unit vectors.

projected length: The length (norm) of the vector projected onto a subspace.

projection: For a given subspace $\mathcal{U} \subset \mathcal{R}^n$ and a point $P \in \mathcal{R}^n$, the point $Q \in \mathcal{U}$ is the "projection" (formally "orthogonal projection") of P onto $\mathcal{U}$ if $\overrightarrow{PQ}$ is perpendicular to $\mathcal{U}$. The mapping of P to Q, which is a linear mapping, is also called the "projection" and is denoted by $\boldsymbol{P}_{\mathcal{U}}$.

projection matrix: The matrix that represents the projection $\boldsymbol{P}_{\mathcal{U}}$ from $\mathcal{R}^n$ onto its subspace $\mathcal{U}$.

rejection: If point $P \in \mathcal{R}^n$ is projected to point $Q \in \mathcal{U}$, the point $P \in \mathcal{R}^n$ is the "rejection" of point Q.

Schmidt orthogonalization: A procedure for converting a set of linearly independent vectors to an orthonormal system. From the already obtained orthonormal set of vectors, we define the projection matrix onto the subspace orthogonal to them, i.e., their orthogonal complement. We apply it to the next vector, normalize it to unit norm, and repeat this. Also called the "Gram–Schmidt orthogonalization."

subspace: A subset of a linear space which is itself a linear space. The subspaces of the 3D space $\mathcal{R}^3$ are the origin O (0D subspace), lines passing through the origin O (1D subspaces), planes passing through the origin O (2D subspaces), and the $\mathcal{R}^3$ itself (3D subspace).

the nD space $\mathcal{R}^n$: The set of points specified by n real coordinates $(x_1, \ldots, x_n)$. Each point is identified with the column vector $\boldsymbol{x} = \left(x_i\right)$, $i = 1, \ldots, n$, with the n coordinates vertically stacked.

- A linear mapping $\boldsymbol{A}$ from $\mathcal{R}^n$ to $\mathcal{R}^m$ is expressed as $\boldsymbol{A} = \sum_{i=1}^{n} \boldsymbol{a}_i \boldsymbol{u}_i^\top$ in terms of an orthogonal basis $\{\boldsymbol{u}_i\}$ and its images $\boldsymbol{a}_i = \boldsymbol{A}\boldsymbol{u}_i$, $i = 1, \ldots, n$ (Eq. (2.1)).

- A linear mapping $\boldsymbol{A}$ from $\mathcal{R}^n$ to $\mathcal{R}^m$ is represented by the matrix $\boldsymbol{A} = \begin{pmatrix} \boldsymbol{a}_1 & \cdots & \boldsymbol{a}_n \end{pmatrix}$ whose columns $\boldsymbol{a}_i$, $i = 1, \ldots, n$, are the image of the natural basis $\{\boldsymbol{e}_i\}$ of $\mathcal{R}^n$ (Eq. (2.3)).

- The projection Q of a point P onto a subspace $\mathcal{U}$ is the point of $\mathcal{U}$ that is the closest to P (Eq. (2.4)).

- A point P is written as the sum of its projection onto a subspace $\mathcal{U}$ and the rejection from it. This defines the direct sum decomposition into the subspace $\mathcal{U}$ and its orthogonal complement (Eq. (2.5)).

- The projection matrix onto the subspace spanned by an orthonormal system $\{\boldsymbol{u}_i\}$, $i = 1, \ldots, r$, is written as $\boldsymbol{P}_{\mathcal{U}} = \sum_{i=1}^{r} \boldsymbol{u}_i \boldsymbol{u}_i^\top$ (Eq. (2.8)).

- The square norm $\|\boldsymbol{x}\|^2$ of vector $\boldsymbol{x}$ equals the sum of the square norm of its projection onto a subspace $\mathcal{U}$ and the square norm of its projection onto its orthogonal complement $\mathcal{U}^\perp$ (Eq. (2.12)).

- The projected length of vector $\boldsymbol{x}$ onto a line is given by the inner product $\langle \boldsymbol{x}, \boldsymbol{u} \rangle$ with the unit direction vector $\boldsymbol{u}$ of the line.

- From a given set of linearly independent vectors $\{\boldsymbol{a}_i\}$, we can produce an orthonormal set $\{\boldsymbol{u}_i\}$ by the Schmidt orthogonalization.

2.7 SUPPLEMENTAL NOTES

What we call "projection," which plays a central role in this book, is formally called "orthogonal projection." However, we do not consider other types of projection in this book, so we simply call it "projection." For interested readers, we briefly mention here the general formulation of the concept of projection.

In general, "projection" is a linear mapping from $\mathcal{R}^n$ to a subspace $\mathcal{U}$ of it. In this chapter, we considered mapping of a point $P \in \mathcal{R}^n$ "orthogonally" onto $\mathcal{U}$, but there exist other possibilities, e.g., mapping P "obliquely" to $\mathcal{U}$ parallel to a given line. Such non-orthogonal projection is formally defined as follows.

Let $\mathcal{U}$ and $\mathcal{V}$ be two subspaces of $\mathcal{R}^n$ with respective dimensions r and $n - r$ for some r. If $\mathcal{U} \cap \mathcal{V} = \{\boldsymbol{0}\}$, it is easy to see that any vector $\boldsymbol{x} \in \mathcal{R}^n$ is uniquely written as

$$\boldsymbol{x} = \boldsymbol{x}_{\mathcal{U}} + \boldsymbol{x}_{\mathcal{V}}, \qquad \boldsymbol{x}_{\mathcal{U}} \in \mathcal{U}, \qquad \boldsymbol{x}_{\mathcal{V}} \in \mathcal{V}. \tag{2.22}$$

($\hookrightarrow$ Problem 2.5.) This is called the *direct sum decomposition* of $\boldsymbol{x}$ to $\mathcal{U}$ and $\mathcal{V}$, and the subspaces $\mathcal{U}$ and $\mathcal{V}$ are said to be *complements* of each other. In particular, if all vectors of $\mathcal{U}$ are orthogonal to all vectors of $\mathcal{V}$, they are *orthogonal complements* of each other, as we saw in Section 2.2.

Given a subspace $\mathcal{U} \subset \mathcal{R}^n$ and its complement $\mathcal{V} \subset \mathcal{R}^n$ and the direct sum decomposition of Eq. (2.22), we say that $\boldsymbol{x}_{\mathcal{U}}$ is the *projection* of $\boldsymbol{x}$ onto $\mathcal{U}$ *along* $\mathcal{V}$. Evidently, the mapping from $\boldsymbol{x}$ to $\boldsymbol{x}_{\mathcal{U}}$ is a linear mapping, and this mapping $\boldsymbol{P}$ is called *projection* (*along* $\mathcal{V}$). If $\mathcal{U}$ and $\mathcal{V}$ are orthogonal complements of each other, $\boldsymbol{P}$ is called *orthogonal projection.*

Let $\boldsymbol{u}_1, \ldots, \boldsymbol{u}_r$ be a (not necessarily orthonormal) basis of $\mathcal{U}$, and $\boldsymbol{u}_{r+1}, \ldots, \boldsymbol{u}_n$ be a (not necessarily orthonormal) basis of $\mathcal{V}$. Then, $\{\boldsymbol{u}_i\}$ is a (not necessarily orthonormal) basis of $\mathcal{R}^n$. By projection $\boldsymbol{P}$ of $\mathcal{R}^n$ onto $\mathcal{U}$ along $\mathcal{V}$, the basis vectors are, by definition, mapped as follows:

$$\boldsymbol{P}\boldsymbol{u}_i = \begin{cases} \boldsymbol{u}_i & i = 1, \ldots, r \\ \boldsymbol{0} & i = r + 1, \ldots, n. \end{cases} \tag{2.23}$$

The matrix representation of $\boldsymbol{P}$ is obtained if we introduce an alternative basis, called the *reciprocal basis*, $\{\boldsymbol{u}_i^\dagger\}$, $i = 1, \ldots, n$, of $\mathcal{R}^n$ defined in such a way that

$$\langle \boldsymbol{u}_i^\dagger, \boldsymbol{u}_j \rangle = \delta_{ij}, \qquad i, j = 1, \ldots, n. \tag{2.24}$$

($\hookrightarrow$ Problem 2.6.) Then, we can see the matrix

$$\boldsymbol{P} = \boldsymbol{u}_1 \boldsymbol{u}_1^{\dagger\top} + \cdots + \boldsymbol{u}_r \boldsymbol{u}_r^{\dagger\top} \tag{2.25}$$

satisfies Eq. (2.23). Note that Eq. (2.24) implies that an orthonormal basis is *self-reciprocal*, i.e., its reciprocal basis is itself. We see that Eq. (2.25) is an extension of Eq. (2.8) to non-orthogonal projection.

We can immediately see that the matrix $\boldsymbol{P}$ of Eq. (2.25) is idempotent (see Eq. (2.14)). In fact,

$$\begin{aligned} \boldsymbol{P}^2 &= \sum_{i=1}^{r} \boldsymbol{u}_i \boldsymbol{u}_i^{\dagger\top} \sum_{j=1}^{r} \boldsymbol{u}_j \boldsymbol{u}_j^{\dagger\top} = \sum_{i,j=1}^{r} \boldsymbol{u}_i \langle \boldsymbol{u}_i^\dagger, \boldsymbol{u}_j \rangle \boldsymbol{u}_j^{\dagger\top} \\ &= \sum_{i,j=1}^{r} \delta_{ij} \boldsymbol{u}_i \boldsymbol{u}_j^{\dagger\top} = \sum_{i=1}^{r} \boldsymbol{u}_i \boldsymbol{u}_i^{\dagger\top} = \boldsymbol{P}. \end{aligned} \tag{2.26}$$

This is also obvious from the geometric meaning of projection: a projected vector is unchanged if projected again. Conversely, an idempotent matrix $\boldsymbol{P}$ represents projection of $\mathcal{R}^n$ onto some subspace $\mathcal{U} \subset \mathcal{R}^n$ along some subspace $\mathcal{V} \subset \mathcal{R}^n$. The subspace $\mathcal{U}$ is the set of all linear combinations of the columns of $\boldsymbol{P}$, called the *columns domain*, which plays an important role in Chapter 4. The subspace $\mathcal{V}$ is the set of all vectors $\boldsymbol{x}$ that are mapped to $\boldsymbol{0}$ ($\boldsymbol{P}\boldsymbol{x} = \boldsymbol{0}$), called the *kernel* of $\boldsymbol{P}$ and denoted by $\ker(\boldsymbol{P})$ ($\hookrightarrow$ Problem 2.7). As shown by Eq. (2.13), an idempotent matrix $\boldsymbol{P}$ represents orthogonal projection if and only if $\boldsymbol{P}$ is symmetric ($\hookrightarrow$ Problem 2.4).

2.8 PROBLEMS

2.1. (1) For an mD vector $\boldsymbol{a} = \left(a_i\right)$ and an nD vector $\boldsymbol{b} = \left(b_i\right)$, which denote vectors whose ith components are a_i and b_i, respectively, show that

$$\boldsymbol{a}\boldsymbol{b}^\top = \left(a_i b_j\right), \tag{2.27}$$

where the right side designates the $m \times n$ matrix whose (i, j) element is $a_i b_j$.

(2) Show that when $m = n$, the identity

$$\mathrm{tr}(\boldsymbol{a}\boldsymbol{b}^\top) = \langle \boldsymbol{a}, \boldsymbol{b} \rangle \tag{2.28}$$

holds, where tr denotes the trace of the matrix.

2.2. Express a point Q of subspace $\mathcal{U}$ in terms of the basis of $\mathcal{U}$, and differentiate the square norm from point P to show that the closest point of $\mathcal{U}$ from P is its projection Q.

2.3. Show that Eqs. (2.13) and (2.14) hold, using Eq. (2.8).

2.4. * Show that a symmetric and idempotent matrix $\boldsymbol{P}$ is the projection matrix onto some subspace.

2.5. Show that if two subspaces $\mathcal{U}$ and $\mathcal{V}$ of $\mathcal{R}^n$ of dimensions r and $n-r$, respectively, for some r such that $\mathcal{U} \cap \mathcal{V} = \{\boldsymbol{0}\}$, any vector $\boldsymbol{x} \in \mathcal{R}^n$ is uniquely written in the form of Eq. (2.22).

2.6. (1) Let $\{\boldsymbol{u}_1, \boldsymbol{u}_2, \boldsymbol{u}_3\}$ is a basis of $\mathcal{R}^3$. Show that its reciprocal basis $\{\boldsymbol{u}_1^\dagger, \boldsymbol{u}_2^\dagger, \boldsymbol{u}_3^\dagger\}$ is given by

$$\boldsymbol{u}_1^\dagger = \frac{\boldsymbol{u}_2 \times \boldsymbol{u}_3}{|\boldsymbol{u}_1, \boldsymbol{u}_2, \boldsymbol{u}_3|}, \qquad \boldsymbol{u}_2^\dagger = \frac{\boldsymbol{u}_3 \times \boldsymbol{u}_1}{|\boldsymbol{u}_1, \boldsymbol{u}_2, \boldsymbol{u}_3|}, \qquad \boldsymbol{u}_3^\dagger = \frac{\boldsymbol{u}_1 \times \boldsymbol{u}_2}{|\boldsymbol{u}_1, \boldsymbol{u}_2, \boldsymbol{u}_3|}, \tag{2.29}$$

where $|\boldsymbol{u}_1, \boldsymbol{u}_2, \boldsymbol{u}_3|$ is the *scalar triple product* of $\boldsymbol{u}_1$, $\boldsymbol{u}_2$, and $\boldsymbol{u}_3$ defined by

$$|\boldsymbol{u}_1, \boldsymbol{u}_2, \boldsymbol{u}_3| = \langle \boldsymbol{u}_1, \boldsymbol{u}_2 \times \boldsymbol{u}_3 \rangle \quad (= \langle \boldsymbol{u}_2, \boldsymbol{u}_3 \times \boldsymbol{u}_1 \rangle = \langle \boldsymbol{u}_3, \boldsymbol{u}_1 \times \boldsymbol{u}_2 \rangle) \tag{2.30}$$

(2) Let $\{\boldsymbol{u}_1, \dots, \boldsymbol{u}_n\}$ be a basis of $\mathcal{R}^n$, and let $\boldsymbol{U} = \begin{pmatrix} \boldsymbol{u}_1 & \cdots & \boldsymbol{u}_n \end{pmatrix}$ (the $n \times n$ matrix consisting of columns $\boldsymbol{u}_1, \dots, \boldsymbol{u}_n$). Show that its reciprocal basis is given by

$$\begin{pmatrix} \boldsymbol{u}_1^\dagger & \cdots & \boldsymbol{u}_n^\dagger \end{pmatrix} = (\boldsymbol{U}^{-1})^\top \quad (= (\boldsymbol{U}^\top)^{-1}). \tag{2.31}$$

2.7. * Show the following for an arbitrary idempotent matrix $\boldsymbol{P}$.

(1) The kernel $\ker(\boldsymbol{P})$ is a subspace spanned by $\{\boldsymbol{e}_i - \boldsymbol{p}_i\}$, $i = 1, \dots, n$, where $\boldsymbol{e}_i$ is the natural basis vector (the ith element is 1 and other elements are 0), and $\boldsymbol{p}_i$ is the ith column of $\boldsymbol{P}$.

(2) The sum $\mathcal{V} + \ker(\boldsymbol{P})$ is a direct sum decomposition of $\mathcal{R}^n$.

(3) For any vector $\boldsymbol{x} \in \mathcal{R}^n$, we obtain a direct sum decomposition

$$\boldsymbol{x} = \boldsymbol{P}\boldsymbol{x} + (\boldsymbol{I} - \boldsymbol{P})\boldsymbol{x}, \qquad \boldsymbol{P}\boldsymbol{x} \in \mathcal{V}, \qquad (\boldsymbol{I} - \boldsymbol{P})\boldsymbol{x} \in \ker(\boldsymbol{P}), \tag{2.32}$$

where $\mathcal{V}$ is the column domain of $\boldsymbol{P}$.

CHAPTER 3

Eigenvalues and Spectral Decomposition

In this chapter, we show that a symmetric matrix can be expressed in terms of its "eigenvalues" and "eigenvectors." This expression is called the "spectral decomposition" of a symmetric matrix. It allows us to convert a symmetric matrix into a diagonal matrix by multiplying it by an "orthogonal matrix" from left and right. This process is called "diagonalization" of a symmetric matrix. We can also express the inverse and powers of a symmetric matrix in terms of its spectral decomposition.

3.1 EIGENVALUES AND EIGENVECTORS

For an $n \times n$ symmetric matrix $\boldsymbol{A}$, there exist n real numbers λ, called the *eigenvalues*, and n nonzero vectors $\boldsymbol{u}$, called the *eigenvectors*, such that

$$\boldsymbol{A}\boldsymbol{u} = \lambda \boldsymbol{u}, \qquad \boldsymbol{u} \neq \boldsymbol{0}. \tag{3.1}$$

($\hookrightarrow$ Appendix A.9.) The n eigenvalues λ_1, ..., λ_n, which may have overlaps, are given as the solution of the nth degree equation

$$\phi(\lambda) \equiv |\lambda \boldsymbol{I} - \boldsymbol{A}| = 0, \tag{3.2}$$

called the *characteristic equation*, where $\boldsymbol{I}$ is the $n \times n$ identity matrix, and $|\cdots|$ denotes the determinant. The nth degree polynomial $\phi(\lambda)$ is called the *characteristic polynomial*. It is known that n eigenvectors $\{\boldsymbol{u}_i\}$, $i = 1, \ldots, n$, can be chosen as an orthonormal system ($\hookrightarrow$ Appendix A.9).

However, we need not actually solve the characteristic equation to obtain eigenvalues and eigenvectors. Various software tools which allow us to compute them with high accuracy and high speed using iterations are available, including the *Jacobi method* and the *Householder method* [18].

3.2 SPECTRAL DECOMPOSITION

Let λ_1, ..., λ_n be the eigenvalues of $\boldsymbol{A}$, and $\{\boldsymbol{u}_i\}$, $i = 1, \ldots, n$, the corresponding orthonormal system of its eigenvectors, which defines an orthonormal basis of $\mathcal{R}$. Equation (3.1) implies that $\boldsymbol{A}$ maps the orthonormal basis vectors $\{\boldsymbol{u}_i\}$ of $\mathcal{R}^n$ to $\lambda_1 \boldsymbol{u}_1$, ..., $\lambda_n \boldsymbol{u}_n$, respectively. Hence, from

Eq. (2.1) the matrix $\boldsymbol{A}$ is written in the form

$$\boldsymbol{A} = \lambda_1 \boldsymbol{u}_1 \boldsymbol{u}_1^\top + \cdots + \lambda_n \boldsymbol{u}_n \boldsymbol{u}_n^\top. \tag{3.3}$$

In other words, *a symmetric matrix can be expressed in terms of its eigenvalues and eigenvectors*. This is called the *spectral decomposition* (this term is used because the "eigenvalue" λ is also called the "spectrum"), or sometimes *eigenvalue decomposition*.

Since each term $\boldsymbol{u}_i \boldsymbol{u}_i^\top$ of Eq. (3.3) is the projection matrix onto the direction, called the *principal axis*, of each eigenvector $\boldsymbol{u}_i$ ($\hookrightarrow$ Eq. (2.15)), Eq. (3.3) expresses the matrix $\boldsymbol{A}$ as a linear combination of the projection matrices onto the principal axes. In other words, the transformation of the space by a symmetric matrix is interpreted to be *projections of each point onto the principal axis directions, followed by multiplication by the respective eigenvalues.*

The identity matrix $\boldsymbol{I}$ maps any orthonormal basis $\{\boldsymbol{u}_i\}$, $i = 1, ..., n$, to itself, i.e., $\boldsymbol{I}\boldsymbol{u}_i = \boldsymbol{u}_i$. Hence, its eigenvalues are all 1, meaning that it has the following spectral decomposition ($\hookrightarrow$ Eq. (2.11)):

$$\boldsymbol{I} = \boldsymbol{u}_1 \boldsymbol{u}_1^\top + \cdots + \boldsymbol{u}_n \boldsymbol{u}_n^\top. \tag{3.4}$$

The number of linearly independent vectors among the n columns of matrix $\boldsymbol{A}$, or the number of linearly independent vectors among its n rows, is called the *rank* of that matrix. Consider an arbitrary linear combination of the columns $\boldsymbol{a}_1, ..., \boldsymbol{a}_n$ of $\boldsymbol{A}$

$$c_1 \boldsymbol{a}_1 + \cdots + c_n \boldsymbol{a}_n = \begin{pmatrix} \boldsymbol{a}_1 & \cdots & \boldsymbol{a}_n \end{pmatrix} \begin{pmatrix} c_1 \\ \vdots \\ c_n \end{pmatrix} = \boldsymbol{A}\boldsymbol{c}, \tag{3.5}$$

where we let $\boldsymbol{c} = \begin{pmatrix} c_i \end{pmatrix}$ (shorthand for a vector whose ith component is c_i). If r of the n eigenvalues are nonzero, we can let $\lambda_{r+1} = \cdots = \lambda_n = 0$ in Eq. (3.3) and write $\boldsymbol{A}\boldsymbol{c}$ as

$$\boldsymbol{A}\boldsymbol{c} = \lambda_1 \boldsymbol{u}_1 \boldsymbol{u}_1^\top \boldsymbol{c} + \cdots + \lambda_r \boldsymbol{u}_r \boldsymbol{u}_r^\top \boldsymbol{c} = \lambda_1 \langle \boldsymbol{u}_1, \boldsymbol{c} \rangle \boldsymbol{u}_1 + \cdots + \lambda_r \langle \boldsymbol{u}_r, \boldsymbol{c} \rangle \boldsymbol{u}_r. \tag{3.6}$$

This means that an arbitrary linear combination of the columns of $\boldsymbol{A}$ is written as a linear combination of mutually orthogonal, hence linearly independent, r vectors $\boldsymbol{u}_1, ..., \boldsymbol{u}_r$ ($\hookrightarrow$ Problem 3.1). Thus, the subspace spanned by $\boldsymbol{a}_1, ..., \boldsymbol{a}_n$ has dimension r, meaning that only r of the n columns are linearly independent. In other words, *the rank r of matrix $\boldsymbol{A}$ equals the number of its nonzero eigenvalues*. Since $\boldsymbol{A}$ is a symmetric matrix, this also holds for the rows, i.e., only r of the n rows are linearly independent.

3.3 DIAGONALIZATION OF SYMMETRIC MATRICES

Equation (3.3) is rewritten as

$$\begin{aligned} A &= \begin{pmatrix} \lambda_1 u_1 & \cdots & \lambda_n u_n \end{pmatrix} \begin{pmatrix} u_1^\top \\ \vdots \\ u_n^\top \end{pmatrix} = \begin{pmatrix} u_1 & \cdots & u_n \end{pmatrix} \begin{pmatrix} \lambda_1 & & \\ & \ddots & \\ & & \lambda_n \end{pmatrix} \begin{pmatrix} u_1^\top \\ \vdots \\ u_n^\top \end{pmatrix} \\ &= U \begin{pmatrix} \lambda_1 & & \\ & \ddots & \\ & & \lambda_n \end{pmatrix} U^\top, \end{aligned} \tag{3.7}$$

where

$$U = \begin{pmatrix} u_1 & \cdots & u_n \end{pmatrix} \tag{3.8}$$

is the matrix consisting of columns $u_1, ..., u_n$ (↪ Problem 3.2). It is easy to see that

$$U^\top U = I \tag{3.9}$$

holds (↪ Problem 3.3). Such a matrix is called an *orthogonal matrix*. If U is an orthogonal matrix, so is its transpose (↪ Problem 3.4). Hence, the rows of an orthogonal matrix are also an orthonormal system. Multiplying Eq. (3.7) by $U^\top$ from left and U from right on both sides, we obtain, using Eq. (3.9), the equality

$$U^\top A U = \begin{pmatrix} \lambda_1 & & \\ & \ddots & \\ & & \lambda_n \end{pmatrix}. \tag{3.10}$$

The left side is called the *congruence transformation* of A by orthogonal matrix U. Thus, *a symmetric matrix A can be diagonalized by the congruence transformation, using the orthogonal matrix U consisting of the unit eigenvectors of A as its rows.* The resulting diagonal elements are the corresponding eigenvalues of A. This process is called the *diagonalization* of a symmetric matrix.

3.4 INVERSE AND POWERS

A matrix is *nonsingular* if its determinant is nonzero. It is well known that a nonsingular matrix A has its *inverse* A^{-1}: $AA^{-1} = A^{-1}A = I$. For an $n \times n$ symmetric matrix, its determinant is the product of all eigenvalues (↪ Problem 3.5). Hence, it is nonsingular if and only if its n eigenvalues are all nonzero, i.e., its rank is n.

Multiplying Eq. (3.1) by A^{-1} on both sides, we obtain $u = \lambda A^{-1} u$, or $A^{-1} u = (1/\lambda) u$. Hence, A^{-1} has the same eigenvectors as A with eigenvalues $1/\lambda$. It follows that A^{-1} has its spectral decomposition

$$A^{-1} = \frac{1}{\lambda_1} u_1 u_1^\top + \cdots + \frac{1}{\lambda_n} u_n u_n^\top. \tag{3.11}$$

($\hookrightarrow$ Problem 3.6). In the same way as Eqs. (3.7) and (3.10), we obtain the relationships

$$A^{-1} = U \begin{pmatrix} 1/\lambda_1 & & \\ & \ddots & \\ & & 1/\lambda_n \end{pmatrix} U^\top, \quad U^\top A^{-1} U = \begin{pmatrix} 1/\lambda_1 & & \\ & \ddots & \\ & & 1/\lambda_n \end{pmatrix}. \tag{3.12}$$

From Eq. (3.1), we see that $A^2u = \lambda Au = \lambda^2 u$, $A^3 u = \lambda^2 Au = \lambda^3 u$, ... , so that $A^N u = \lambda^N u$. Hence, for an arbitrary natural number N, the matrix A^N has the same eigenvectors as A with eigenvalues λ^N. It follows that it has the spectral decomposition

$$A^N = \lambda_1^N u_1 u_1^\top + \cdots + \lambda_n^N u_n u_n^\top. \tag{3.13}$$

From this, we obtain, as in the case of Eq. (3.12), the expressions

$$A^N = U \begin{pmatrix} \lambda_1^N & & \\ & \ddots & \\ & & \lambda_n^N \end{pmatrix} U^\top, \quad U^\top A^N U = \begin{pmatrix} \lambda_1^N & & \\ & \ddots & \\ & & \lambda_n^N \end{pmatrix}. \tag{3.14}$$

It is easy to see that this also applies to an arbitrary polynomial $f(x)$ so that we obtain

$$f(A) = f(\lambda_1) u_1 u_1^\top + \cdots + f(\lambda_n) u_n u_n^\top, \tag{3.15}$$

$$f(A) = U \begin{pmatrix} f(\lambda_1) & & \\ & \ddots & \\ & & f(\lambda_n) \end{pmatrix} U^\top, \quad U^\top f(A) U = \begin{pmatrix} f(\lambda_1) & & \\ & \ddots & \\ & & f(\lambda_n) \end{pmatrix}. \tag{3.16}$$

These equations can be extended to an arbitrary function $f(x)$ for which its power series expansion converges. Furthermore, for any function $f(x)$ for which $f(\lambda_i)$, $i = 1, ..., n$, are defined, we can "define" $f(A)$ via Eq. (3.15). Suppose all the eigenvalues of A is nonnegative (such a matrix is said to be *positive semidefinite*; it is *positive definite* if al the eigenvalues are positive). Then, we can define the "square root" $\sqrt{A}$ of A by

$$\sqrt{A} = \sqrt{\lambda_1} u_1 u_1^\top + \cdots + \sqrt{\lambda_n} u_n u_n^\top, \tag{3.17}$$

$$\sqrt{A} = U \begin{pmatrix} \sqrt{\lambda_1} & & \\ & \ddots & \\ & & \sqrt{\lambda_n} \end{pmatrix} U^\top, \quad U^\top \sqrt{A} U = \begin{pmatrix} \sqrt{\lambda_1} & & \\ & \ddots & \\ & & \sqrt{\lambda_n} \end{pmatrix}. \tag{3.18}$$

($\hookrightarrow$ Problem 3.7).

We can view Eqs. (3.4) and (3.11) as special cases of Eq. (3.13) with $N = 0, -1$, where we define $A^0 = I$. For a nonsingular matrix A, we can write $A^{-N} = (A^{-1})^N$ $(= (A^N)^{-1})$ for a natural number N ($\hookrightarrow$ Problem 3.8). Combining Eqs. (3.11) and (3.13), we see that Eq. (3.13) holds for an arbitrary integer N. If A is a positive definite symmetric matrix, N can be extended to an arbitrary real number.

3.5 GLOSSARY AND SUMMARY

characteristic equation: The nth degree equation $\phi(\lambda) = 0$ of λ obtained by letting the characteristic polynomial be 0. All eigenvalues of the matrix $\boldsymbol{A}$ are the roots of the characteristic equation.

characteristic polynomial: The polynomial of degree n in variable λ defined by $\phi(\lambda) = |\lambda \boldsymbol{I} - \boldsymbol{A}|$, where $\boldsymbol{I}$ is the $n \times n$ identity matrix and $| \cdots |$ designates the determinant.

congruence transformation: Sandwiching an $n \times n$ symmetric matrix $\boldsymbol{A}$ by an orthogonal matrix $\boldsymbol{U}$ in the form of $\boldsymbol{U}^\top \boldsymbol{A} \boldsymbol{U}$.

diagonalization: Converting a symmetric matrix $\boldsymbol{A}$ into a diagonal matrix consisting of its eigenvalues as the diagonal elements. This is done by the congruence transformation using the orthogonal matrix consisting of its unit eigenvectors as the columns.

eigenvalue: A (real or complex) number λ such that $\boldsymbol{A}\boldsymbol{u} = \lambda \boldsymbol{u}$, $\boldsymbol{u} \neq \boldsymbol{0}$, holds for an $n \times n$ matrix $\boldsymbol{A}$; the vector $\boldsymbol{u}$ is called the "eigenvector" of $\boldsymbol{A}$ for the eigenvalue λ.

nonsingular matrix: An $n \times n$ matrix $\boldsymbol{A}$ is nonsingular if its determinant $|\boldsymbol{A}|$ is nonzero. For a symmetric matrix, this is equivalent to say that its eigenvalues are all nonzero. The columns and rows of a nonsingular matrix $\boldsymbol{A}$ are linearly independent, i.e., the rank is n, and the inverse $\boldsymbol{A}^{-1}$ exists.

orthogonal matrix: An $n \times n$ matrix whose columns are an orthonormal system. Its rows are also an orthonormal system.

positive definite symmetric matrix: A symmetric matrix whose eigenvalues are all positive. If the eigenvalues are positive or zero, it is a "positive semidefinite symmetric matrix."

principal axis: The directions of the eigenvectors $\boldsymbol{u}_i$, $i = 1, ..., n$, of a symmetric matrix.

rank: The number of linearly independent columns of a matrix (= the number of its independent rows). For a symmetric matrix, it equals the number of its nonzero eigenvalues.

spectral decomposition: The expression of an $n \times n$ symmetric matrix $\boldsymbol{A}$ in terms of its eigenvalues λ_i and eigenvectors $\boldsymbol{u}_i$, $i = 1, ..., n$, in the form $\boldsymbol{A} = \sum_{i=1}^{n} \lambda_i \boldsymbol{u}_i \boldsymbol{u}_i^\top$.

- All eigenvalues λ of a symmetric matrix are real, and their eigenvectors $\boldsymbol{u}$ consist of real components.
- Software tools are available for computing eigenvalues and eigenvectors.
- Eigenvectors for different eigenvalues are orthogonal to each other.

- The eigenvectors $\{\boldsymbol{u}_i\}$, $i = 1, ..., n$, of an $n \times n$ symmetric matrix define an orthonormal basis of $\mathcal{R}^n$.
- An $n \times n$ symmetric matrix is expressed in terms of its eigenvalues λ_i and eigenvectors $\boldsymbol{u}_i$, $i = 1, ..., n$, in the form $\boldsymbol{A} = \sum_{i=1}^{n} \lambda_i \boldsymbol{u}_i \boldsymbol{u}_i^\top$ ("spectral decomposition").
- The spectral decomposition of a symmetric matrix $\boldsymbol{A}$ is written as $\boldsymbol{A} = \boldsymbol{U}\boldsymbol{\Lambda}\boldsymbol{U}^\top$, where $\boldsymbol{U}$ is a matrix consisting of the eigenvectors as its columns and $\boldsymbol{\Lambda}$ is a diagonal matrix with the eigenvalues as the diagonal elements.
- The number of nonzero eigenvalues of a symmetric matrix $\boldsymbol{A}$ is equal to the number of linearly independent columns of $\boldsymbol{A}$ and also to the number of its linearly independent rows (= the "rank" of the matrix $\boldsymbol{A}$).
- A symmetric matrix $\boldsymbol{A}$ is converted to a diagonal matrix consisting of its eigenvalues as the diagonal elements by a congruence transformation $\boldsymbol{U}^\top \boldsymbol{A} \boldsymbol{U}$, using the orthogonal matrix $\boldsymbol{U}$ consisting of the unit eigenvectors of $\boldsymbol{A}$ ("diagonalization" of a symmetric matrix).
- For a nonsingular symmetric matrix $\boldsymbol{A}$ having the spectral decomposition $\boldsymbol{A} = \sum_{i=1}^{n} \lambda_i \boldsymbol{u}_i \boldsymbol{u}_i^\top$, its inverse is given by $\boldsymbol{A}^{-1} = \sum_{i=1}^{n} (1/\lambda_i) \boldsymbol{u}_i \boldsymbol{u}_i^\top$.
- For a positive definite symmetric matrix $\boldsymbol{A}$ having the spectral decomposition $\boldsymbol{A} = \sum_{i=1}^{n} \lambda_i \boldsymbol{u}_i \boldsymbol{u}_i^\top$, its Nth power is given by $\boldsymbol{A}^N = \sum_{i=1}^{n} \lambda_i^N \boldsymbol{u}_i \boldsymbol{u}_i^\top$, where N is an arbitrary real number, which can be negative or non-integer.

3.6 SUPPLEMENTAL NOTES

In this chapter, we only considered symmetric matrices and characterized them in terms of their eigenvalues and eigenvectors. There is a good reason for that. In fact, if a matrix $\boldsymbol{A}$ is not symmetric or even if it is a rectangular matrix, the self-products $\boldsymbol{A}\boldsymbol{A}^\top$ and $\boldsymbol{A}^\top\boldsymbol{A}$ are symmetric matrices, and the matrix $\boldsymbol{A}$ is characterized in terms of their eigenvalues and eigenvectors. This is the theme of the next chapter.

On the other hand, eigenvalues and eigenvectors are also defined for non-symmetric matrices by Eq. (3.1). For non-symmetric matrices, however, eigenvalues may not be real, and their eigenvectors may not have real components. In order to analyze the properties of eigenvalues and eigenvectors, we need to consider $\mathcal{C}^n$, i.e., the set of points with n complex coordinates $(x_1, ..., x_n)$. The properties of eigenvalues and eigenvectors in $\mathcal{C}^n$ have been extensively studied in the past, in particular in physics in relation to quantum mechanics. Here, we briefly summarize the classical results about eigenvalues and eigenvectors in $\mathcal{C}^n$.

By the same reasoning as in the case of real symmetric matrices (↪ Appendix A.9), the eigenvalues of $\boldsymbol{A}$ are roots of the characteristic polynomial $\phi(\lambda)$ defined by Eq. (3.2). Hence, an

$n \times n$ matrix $\boldsymbol{A}$ has n eigenvalues in the complex domain with possible multiplicities. It is easy to show that eigenvectors for different eigenvalues are linearly independent ($\hookrightarrow$ Problem 3.9). For eigenvalues with mulitplicities, however, we need careful analysis. Suppose the characteristic polynomial has the form of $\phi(\lambda) = (\lambda - \lambda_1)^{m_1} \cdots (\lambda - \lambda_r)^{m_r}$, having distinct eigenvalues λ_1, ..., λ_r with respective multiplicities m_1, ..., m_r, $(m_1 + \cdots + m_r = n)$. From Eq. (3.1), eigenvectors $\boldsymbol{u}$ for eigenvalue λ_i satisfy

$$(\boldsymbol{A} - \lambda_i \boldsymbol{I})\boldsymbol{u} = \boldsymbol{0}. \tag{3.19}$$

If the rank r_i of the matrix $\boldsymbol{A} - \lambda_i \boldsymbol{I}$ is $n - m_i$, we can find m_i linearly independent eigenvectors. Hence, if all eigenvalues are simple roots ($m_i = 1$), we can find n ($= \sum_{i=1}^{r} m_i$) linearly independent eigenvectors. However, if $r_i < n - m_i$ for some eigenvalues, we may not be able to find n linearly independent eigenvectors.

It can be shown (we omit the proof) that an $n \times n$ matrix $\boldsymbol{A}$ has n linearly independent eigenvectors if and only if it satisfies

$$\boldsymbol{A}\boldsymbol{A}^* = \boldsymbol{A}^*\boldsymbol{A}, \tag{3.20}$$

where $\boldsymbol{A}^* = \bar{\boldsymbol{A}}^\top$ (transpose of complex conjugate); $\boldsymbol{A}^*$ is called the *adjoint matrix* of $\boldsymbol{A}$. A matrix $\boldsymbol{A}$ is said to be *normal* if Eq. (3.20) is satisfied. Evidently, a real symmetric matrix $\boldsymbol{A}$ ($\boldsymbol{A}^* = \boldsymbol{A}^\top$ $= \boldsymbol{A}$) is a normal matrix. An orthogonal matrix $\boldsymbol{U}$ is also a normal matrix, since $\boldsymbol{U}^\top\boldsymbol{U} = \boldsymbol{U}\boldsymbol{U}^\top$ $= \boldsymbol{I}$ ($\hookrightarrow$ Problems 3.3 and 3.4).

For an $n \times n$ normal matrix $\boldsymbol{A}$ having eigenvalues λ_1, ..., λ_n (with possible multiplicities), we can find linearly independent eigenvectors $\boldsymbol{u}_1$, ..., $\boldsymbol{u}_n$, which form a basis of of $\mathcal{C}^n$. Then, we can define the *reciprocal basis* $\{\boldsymbol{u}^\dagger\}$, $i = 1, \ldots, n$, that satisfies Eq. (2.24). It is easy to see that, just as in $\mathcal{R}^n$, the reciprocal basis is given by Eq. (2.31). We can also see that $\boldsymbol{A}$ is expressed in the form

$$\boldsymbol{A} = \lambda_1 \boldsymbol{u}_1 \boldsymbol{u}_1^{\dagger\top} + \cdots + \lambda_n \boldsymbol{u}_n \boldsymbol{u}_n^{\dagger\top}. \tag{3.21}$$

In fact, we can see from Eq. (2.24) that $\boldsymbol{A}\boldsymbol{u}_i = \lambda_i \boldsymbol{u}_i$, $i = 1, \ldots, n$, holds. We see that this is the extension of the spectral decomposition of Eq. (3.3) for a symmetric matrix in $\mathcal{R}^n$ to a normal matrix in $\mathcal{C}^n$ ($\hookrightarrow$ Eq. (2.25)).

As we did in Eq. (3.7), we can rewrite Eq. (3.21) as

$$\begin{aligned} \boldsymbol{A} &= \begin{pmatrix} \lambda_1 \boldsymbol{u}_1 & \cdots & \lambda_n \boldsymbol{u}_n \end{pmatrix} \begin{pmatrix} \boldsymbol{u}_1^{\dagger\top} \\ \vdots \\ \boldsymbol{u}_n^{\dagger\top} \end{pmatrix} = \begin{pmatrix} \boldsymbol{u}_1 & \cdots & \boldsymbol{u}_n \end{pmatrix} \begin{pmatrix} \lambda_1 & & \\ & \ddots & \\ & & \lambda_n \end{pmatrix} \begin{pmatrix} \boldsymbol{u}_1^{\dagger\top} \\ \vdots \\ \boldsymbol{u}_n^{\dagger\top} \end{pmatrix} \\ &= \boldsymbol{U} \begin{pmatrix} \lambda_1 & & \\ & \ddots & \\ & & \lambda_n \end{pmatrix} \boldsymbol{U}^{-1}, \end{aligned} \tag{3.22}$$

where we have noted Eq. (2.31) of the definition of $\{\boldsymbol{u}_i^\dagger\}$, $i = 1, ..., n$. Multiplying Eq. (3.22) by $\boldsymbol{U}^{-1}$ from left and $\boldsymbol{U}$ from right on both sides, we obtain

$$\boldsymbol{U}^{-1}\boldsymbol{A}\boldsymbol{U} = \begin{pmatrix} \lambda_1 & & \\ & \ddots & \\ & & \lambda_n \end{pmatrix}. \tag{3.23}$$

This is the extension of the diagonalization of a symmetric matrix in $\mathcal{R}^n$ given by Eq. (3.10) to a normal matrix in $\mathcal{C}^n$. The left side is called the *similarity transformation* of $\boldsymbol{A}$ by nonsingular matrix $\boldsymbol{U}$. Thus, *a normal matrix* $\boldsymbol{A}$ *can be diagonalized in* $\mathcal{C}^n$ *by the similarity transformation, using the nonsingular matrix* $\boldsymbol{U}$ *consisting of the eigenvectors of* $\boldsymbol{A}$ *as the columns*. The resulting diagonal elements are the corresponding eigenvalues of $\boldsymbol{A}$.

For a non-normal matrix $\boldsymbol{A}$, however, we cannot find a nonsingular matrix $\boldsymbol{U}$ such that Eq. (3.23) holds. Nevertheless, we can find a nonsingular matrix $\boldsymbol{U}$ that makes the right side of Eq. (3.23) "block-diagonal." Let λ_i, ..., λ_r be distinct eigenvalues of $\boldsymbol{A}$, each with multiplicity m_i ($\sum_{i=1}^r m_i = n$), and let $\boldsymbol{u}_1$, ..., $\boldsymbol{u}_r$ be their corresponding eigenvectors (they are not unique but linearly independent ↪ Problem 3.9). If $r = n$, the $n \times n$ matrix $\boldsymbol{U} = \begin{pmatrix} \boldsymbol{u}_1 & \cdots & \boldsymbol{u}_n \end{pmatrix}$ having them as columns is a nonsingular matrix, and Eq. (3.23) holds. Suppose $r < n$. For each $\boldsymbol{u}_i$, we want to find $\boldsymbol{u}_i^{(j)}$, $j = 1, ..., m_i$, such that $\boldsymbol{u}_1^{(1)}$, ..., $\boldsymbol{u}_1^{(m_1)}$, $\boldsymbol{u}_2^{(1)}$, ..., $\boldsymbol{u}_2^{(m_2)}$, ..., $\boldsymbol{u}_r^{(m_r)}$ are linearly independent. They may not satisfy Eq. (3.19).

We say that if a vector $\boldsymbol{u}$ satisfies

$$(\boldsymbol{A} - \lambda \boldsymbol{I})^k \boldsymbol{u} = \boldsymbol{0} \tag{3.24}$$

for some integer k (≥ 1), the vector $\boldsymbol{u}$ is a *generalized eigenvector* of $\boldsymbol{A}$ for eigenvalue λ (if $k = 1$, it is the usual eigenvector). It is known (we omit the proof) that for each i we can find m_i linearly independent generalized vectors $\boldsymbol{u}_i^{(1)}$, ..., $\boldsymbol{u}_i^{(m_i)}$ of $\boldsymbol{A}$. We can show (we omit the proof) that the similarity transformation of $\boldsymbol{A}$ using the $n \times n$ matrix

$$\boldsymbol{U} = \begin{pmatrix} \boldsymbol{u}_1^{(1)} & \cdots & \boldsymbol{u}_1^{(m_1)} & \boldsymbol{u}_2^{(1)} & \cdots & \boldsymbol{u}_2^{(m_2)} & \cdots & \boldsymbol{u}_r^{(m_r)} \end{pmatrix} \tag{3.25}$$

has the block-diagonal form

$$\boldsymbol{U}^{-1}\boldsymbol{A}\boldsymbol{U} = \begin{pmatrix} \boldsymbol{J}_1 & & \\ & \ddots & \\ & & \boldsymbol{J}_r \end{pmatrix}, \tag{3.26}$$

where $\boldsymbol{J}_i$ is an $m_i \times m_i$ matrix. The form of each block $\boldsymbol{J}_i$ depends on the choice of the generalized eigenvectors for λ_i, but it is known (we omit the derivation, but you may guess how it goes from Problem 3.10) that we can choose them so that each $\boldsymbol{J}_i$ has a block-diagonal form

$$\boldsymbol{J}_i = \begin{pmatrix} \boldsymbol{J}_i^{(1)} & & \\ & \ddots & \\ & & \boldsymbol{J}_i^{(s)} \end{pmatrix}, \tag{3.27}$$

for some s, where each $\boldsymbol{J}_i^{(l)}$, $l = 1, ..., s$, has the form

$$\boldsymbol{J}_i^{(l)} = \begin{pmatrix} \lambda_i & 1 & & & \\ & \lambda_i & 1 & & \\ & & \ddots & \ddots & \\ & & & & 1 \\ & & & & \lambda_i \end{pmatrix}. \tag{3.28}$$

This form is called the *Jordan cell* for eigenvalue λ_i. Thus, *we can choose generalized eigenvectors so that the similarity transformation of Eq. (3.26) is a block-diagonal matrix consisting of Jordan cells of the from of Eq. (3.28)*. The resulting matrix is called the *Jordan canonical form*.

The Jordan canonical form does not usually appear in pattern information processing. This is partly because matrices involved are mostly symmetric and partly because data are obtained by "sensing," including camera imaging. The Jordan canonical form describes the structure of the "degeneracy" of eigenvalues. If all eigenvalues are distinct, we obtain diagonal canonical forms ($\hookrightarrow$ Problem 3.9). Sensor data generally contain noise, so the characteristic polynomial usually have no multiple roots for sensor data.

However, we frequently encounter the Jordan canonical form in control theory, in particular when high-order differentiation is involved. This is because we "design" a control system that way. In particular, the multiplicities m_i of eigenvalues λ_i are "parameters" of the designed system, and we analyze and characterize the system behavior, using the knowledge that the system has a specified type of degeneracies. In short, the Jordan canonical form does not come out of numerical data but rather it appears by design.

For pattern information processing, non-symmetric and rectangular matrices are most conveniently treated by the singular value decomposition described in the next chapter.

3.7 PROBLEMS

3.1. Show that mutually orthogonal nonzero vectors $\boldsymbol{u}_1, ..., \boldsymbol{u}_m$ are linearly independent.

3.2. Show that for nD vectors $\boldsymbol{a}_1, ..., \boldsymbol{a}_m$ and $\boldsymbol{b}_1, ..., \boldsymbol{b}_m$ the following relationship holds:

$$\sum_{i=1}^{m} \boldsymbol{a}_i \boldsymbol{b}_i^\top = \begin{pmatrix} \boldsymbol{a}_1 & \cdots & \boldsymbol{a}_m \end{pmatrix} \begin{pmatrix} \boldsymbol{b}_1^\top \\ \vdots \\ \boldsymbol{b}_m^\top \end{pmatrix} = \boldsymbol{A}\boldsymbol{B}^\top. \tag{3.29}$$

Here, $\boldsymbol{A}$ and $\boldsymbol{B}$ are $n \times m$ matrices having columns $\boldsymbol{a}_1, \ldots, \boldsymbol{a}_m$ and columns $\boldsymbol{b}_1, \ldots, \boldsymbol{b}_m$, respectively.

3.3. Show that the matrix $\boldsymbol{U}$ of Eq. (3.8) is orthogonal, i.e., its columns form an orthonormal system, if and only if Eq. (3.9) holds.

3.4. Show that if $\boldsymbol{U}$ is an orthogonal matrix, so is $\boldsymbol{U}^\top$, i.e., an orthogonal has not only orthonormal columns but also orthonormal rows.

3.5. (1) Show that orthogonal matrices have determinant ± 1.

(2) Show that the determinant $|\boldsymbol{A}|$ of a symmetric matrix $\boldsymbol{A}$ equals the product of its all eigenvalues.

3.6. Show that the matrix $\boldsymbol{A}$ of Eq. (3.3) and the matrix $\boldsymbol{A}^{-1}$ of Eq. (3.11) satisfy $\boldsymbol{A}^{-1}\boldsymbol{A} = \boldsymbol{I}$ by computing their product.

3.7. For the matrix $\sqrt{\boldsymbol{A}}$ defined by Eq. (3.17), or by the first equation of Eq. (3.18), show that $(\sqrt{\boldsymbol{A}})^2 = \boldsymbol{A}$ holds.

3.8. Show that for a nonsingular matrix $\boldsymbol{A}$, the identity

$$(\boldsymbol{A}^{-1})^N = (\boldsymbol{A}^N)^{-1} \tag{3.30}$$

holds for any natural number N.

3.9.* Let $\boldsymbol{u}_1, \ldots, \boldsymbol{u}_m$ be the eigenvectors of $\boldsymbol{A}$ that have distinct eigenvalues $\lambda_1, \ldots, \lambda_m$ in $\mathcal{C}^n$. Show that $\boldsymbol{u}_1, \ldots, \boldsymbol{u}_m$ are linearly independent.

3.10.* Suppose $\boldsymbol{A}$ has a generalized eigenvector $\boldsymbol{u}_n$ for eigenvalue λ such that

$$(\boldsymbol{A} - \lambda \boldsymbol{I})^n \boldsymbol{u} = \boldsymbol{0} \tag{3.31}$$

and that

$$(\boldsymbol{A} - \lambda \boldsymbol{I})^k \boldsymbol{u} \neq \boldsymbol{0}, \qquad k < n. \tag{3.32}$$

Define $\boldsymbol{u}_1, \ldots, \boldsymbol{u}_{n-1}$ by

$$\begin{aligned} \boldsymbol{u}_1 &= (\boldsymbol{A} - \lambda \boldsymbol{I})^{n-1} \boldsymbol{u}_n, \\ \boldsymbol{u}_2 &= (\boldsymbol{A} - \lambda \boldsymbol{I})^{n-2} \boldsymbol{u}_n, \\ &\cdots \\ \boldsymbol{u}_{n-1} &= (\boldsymbol{A} - \lambda \boldsymbol{I}) \boldsymbol{u}_n. \end{aligned} \tag{3.33}$$

Let $\boldsymbol{U} = \begin{pmatrix} \boldsymbol{u}_1 & \cdots & \boldsymbol{u}_n \end{pmatrix}$ be a matrix having them as columns. This is a nonsingular matrix (we omit the proof). Show that the similarity transformation using this $\boldsymbol{U}$ reduces $\boldsymbol{A}$ to the form of

$$\boldsymbol{U}^{-1}\boldsymbol{A}\boldsymbol{U} = \begin{pmatrix} \lambda & 1 & & & \\ & \lambda & 1 & & \\ & & \ddots & \ddots & \\ & & & \ddots & 1 \\ & & & & \lambda \end{pmatrix}. \tag{3.34}$$

CHAPTER 4

Singular Values and Singular Value Decomposition

The spectral decomposition described in the preceding chapter is defined only for symmetric matrices, hence for square matrices. In this chapter, we extend it to arbitrary rectangular matrices; we define the "singular value decomposition," which expresses any matrix in terms of its "singular values" and "singular vectors." The singular vectors form a basis of the subspace spanned by the columns or the rows, defining a projection matrix onto it.

4.1 SINGULAR VALUES AND SINGULAR VECTORS

For an $m \times n$ matrix $\boldsymbol{A}$ which is not the zero matrix $\boldsymbol{O}$, i.e., the matrix whose elements are all zero, there exist a positive number σ (> 0), an mD vector $\boldsymbol{u}$ ($\neq \boldsymbol{0}$), and an nD vector $\boldsymbol{v}$ ($\neq \boldsymbol{0}$) such that

$$\boldsymbol{A}\boldsymbol{v} = \sigma \boldsymbol{u}, \qquad \boldsymbol{A}^\top \boldsymbol{u} = \sigma \boldsymbol{v}, \qquad \sigma > 0, \quad \boldsymbol{u} \neq \boldsymbol{0}, \quad \boldsymbol{v} \neq \boldsymbol{0}. \tag{4.1}$$

The number σ and the vectors $\boldsymbol{u}$ and $\boldsymbol{v}$ are called the *singular value*, the *left singular vector*, and the *right singular vector*, respectively. The left and right singular vectors $\boldsymbol{u}$ and $\boldsymbol{v}$ are simply called the *singular vectors*. There exist r set of such singular values and singular vectors, where r is the rank of the matrix $\boldsymbol{A}$, i.e., the number of linearly independent columns and the number of linearly independent rows (we discuss this shortly).

Multiplying the second equation of Eq. (4.1) by $\boldsymbol{A}$ from left on both sides, and multiplying the first equation by $\boldsymbol{A}^\top$ from left on both sides, we see that

$$\boldsymbol{A}\boldsymbol{A}^\top \boldsymbol{u} = \sigma^2 \boldsymbol{u}, \qquad \boldsymbol{A}^\top \boldsymbol{A}\boldsymbol{v} = \sigma^2 \boldsymbol{v}. \tag{4.2}$$

Namely, the left singular vector $\boldsymbol{u}$ is the eigenvector of the $m \times m$ symmetric matrix $\boldsymbol{A}\boldsymbol{A}^\top$, and the right singular vector $\boldsymbol{v}$ is the eigenvector of the $n \times n$ symmetric matrix $\boldsymbol{A}^\top\boldsymbol{A}$. The squared singular value σ^2 is the eigenvalue of both of them ($\hookrightarrow$ Problem 4.1). It is easy to see that $\boldsymbol{A}\boldsymbol{A}^\top$ and $\boldsymbol{A}^\top\boldsymbol{A}$ have a common positive eigenvalue σ^2 and that their eigenvectors $\boldsymbol{u}$ and $\boldsymbol{v}$ are related by Eq. (4.1) ($\hookrightarrow$ Problem 4.2).

Let $\sigma_1 \geq \cdots \geq \sigma_r$ (> 0) be the singular values of $\boldsymbol{A}$, where some of them may overlap. Since the corresponding r left singular vectors $\boldsymbol{u}_1$, ..., $\boldsymbol{u}_r$ and r right singular vectors are both eigenvectors of symmetric matrices, they can be chosen to form orthonormal systems.

For actually computing the singular values and singular vectors, we need not compute the eigenvalues and eigenvectors of $\boldsymbol{A}\boldsymbol{A}^\top$ and $\boldsymbol{A}^\top\boldsymbol{A}$. Various software tools that can compute them with high speed and high accuracy are available. A typical one consists of transformation to a *bidiagonal matrix* by means of the Householder method and application of the *Golub–Reinsch method*.

4.2 SINGULAR VALUE DECOMPOSITION

An $m \times n$ matrix $\boldsymbol{A}$ defines a linear mapping from the nD space $\mathcal{R}^n$ to the mD space $\mathcal{R}^m$ ($\hookrightarrow$ Appendix A.1). We can extend the orthonormal system $\boldsymbol{u}_1, ..., \boldsymbol{u}_r$ of the r left singular vectors to an orthonormal basis $\{\boldsymbol{u}_1, ..., \boldsymbol{u}_r, \boldsymbol{u}_{r+1}, ..., \boldsymbol{u}_m\}$ of $\mathcal{R}^m$, say, by the Schmidt orthogonalization ($\hookrightarrow$ Section 2.4). Similarly, we can extend the orthonormal system $\boldsymbol{v}_1, ..., \boldsymbol{v}_r$ of the r right singular vectors to an orthonormal basis $\{\boldsymbol{v}_1, ..., \boldsymbol{v}_r, \boldsymbol{v}_{r+1}, ..., \boldsymbol{v}_n\}$ of $\mathcal{R}^n$. From Eq (4.2), these are eigenvectors of $\boldsymbol{A}\boldsymbol{A}^\top$ and $\boldsymbol{A}^\top\boldsymbol{A}$, and the eigenvalues for $\boldsymbol{u}_{r+1}, ..., \boldsymbol{u}_m$ and $\boldsymbol{v}_{r+1}, ..., \boldsymbol{v}_n$ are all 0:

$$\begin{aligned} \boldsymbol{A}\boldsymbol{A}^\top\boldsymbol{u}_i &= \boldsymbol{0}, \qquad i = r+1, ..., m, \\ \boldsymbol{A}^\top\boldsymbol{A}\boldsymbol{v}_i &= \boldsymbol{0}, \qquad i = r+1, ..., n. \end{aligned} \tag{4.3}$$

From the second equation, we see that $\boldsymbol{A}\boldsymbol{v}_i = \boldsymbol{0}$, $i = r+1, ..., n$ ($\hookrightarrow$ Problem 4.3(1)). This and the first equation of Eq. (4.1) imply that $\boldsymbol{A}$ maps the orthonormal basis vectors $\{\boldsymbol{v}_1, ..., \boldsymbol{v}_n\}$ of $\mathcal{R}^n$ to $\sigma_1\boldsymbol{u}_1, ..., \sigma_r\boldsymbol{u}_r, \boldsymbol{0}, ..., \boldsymbol{0}$, respectively. Hence, from Eq. (2.1), we see that $\boldsymbol{A}$ is expressed as

$$\boldsymbol{A} = \sigma_1\boldsymbol{u}_1\boldsymbol{v}_1^\top + \cdots + \sigma_r\boldsymbol{u}_r\boldsymbol{v}_r^\top, \qquad \sigma_1 \geq \cdots \geq \sigma_r > 0. \tag{4.4}$$

Similarly, we see from Eq. (4.3) that $\boldsymbol{A}^\top\boldsymbol{u}_i = \boldsymbol{0}$, $i = r+1, ..., m$ ($\hookrightarrow$ Problem 4.3(2)). This and the second equation of Eq. (4.1) imply that $\boldsymbol{A}^\top$ maps the orthonormal basis vectors $\{\boldsymbol{u}_1, ..., \boldsymbol{u}_n\}$ of $\mathcal{R}^m$ to $\sigma_1\boldsymbol{v}_1, ..., \sigma_r\boldsymbol{v}_r, \boldsymbol{0}, ..., \boldsymbol{0}$, respectively. Hence, from Eq. (2.1), we see that $\boldsymbol{A}^\top$ is expressed as

$$\boldsymbol{A}^\top = \sigma_1\boldsymbol{v}_1\boldsymbol{u}_1^\top + \cdots + \sigma_r\boldsymbol{v}_r\boldsymbol{u}_r^\top, \qquad \sigma_1 \geq \cdots \geq \sigma_r > 0, \tag{4.5}$$

which is the transpose of Eq. (4.4) on both sides. Thus, *an arbitrary matrix is expressed in terms of its singular values and singular vectors*. This is called the *singular value decomposition*.

4.3 COLUMN DOMAIN AND ROW DOMAIN

Let $\mathcal{U}$ be the subspace spanned by the n columns of $\boldsymbol{A}$, and $\mathcal{V}$ the subspace spanned by its m rows. We call the the *column domain* and the *row domain*, respectively.

Consider an arbitrary linear combination of the columns $\boldsymbol{a}_1,..., \boldsymbol{a}_n$ of $\boldsymbol{A}$

$$c_1\boldsymbol{a}_1 + \cdots + c_n\boldsymbol{a}_n = \begin{pmatrix} \boldsymbol{a}_1 & \cdots & \boldsymbol{a}_n \end{pmatrix}\begin{pmatrix} c_1 \\ \vdots \\ c_n \end{pmatrix} = \boldsymbol{A}\boldsymbol{c}, \tag{4.6}$$

where we let $c = \left(c_i\right)$. From Eq. (4.4), this is rewritten as follows:

$$Ac = \sigma_1 u_1 v_1^\top c + \cdots + \sigma_r u_r v_r^\top c = \sigma_1 \langle v_1, c\rangle u_1 + \cdots + \sigma_r \langle v_r, c\rangle u_r. \tag{4.7}$$

Namely, an arbitrary linear combination of the columns of A is a linear combination of mutually orthogonal, hence linearly independent (↪ Problem 4.4), vectors u_1, ..., u_r. Thus, the column domain $\mathcal{U}$ spanned by a_1, ..., a_n is an rD subspace, for which u_1, ..., u_r are an orthonormal basis. It follows that *only r columns are linearly independent*.

The rows of A are the columns of $A^\top$. Hence, from Eq. (4.5) an arbitrary linear combination of rows is expressed as a linear combination of v_1, ..., v_r. Thus, the row domain $\mathcal{V}$ spanned by rows of A is an rD subspace, for which v_1, ..., v_r are an orthonormal basis. It follows that *only r rows are linearly independent*.

From these, we conclude that *the rank r of A equals the number of the singular values of A* and that *the left singular vectors $\{u_i\}$, $i = 1, ..., r$, and the right singular vectors $\{v_i\}$, $i = 1, ..., r$, constitute the orthonormal bases of the columns domain $\mathcal{U}$ and the row domain $\mathcal{V}$, respectively*.

From Eq. (2.8), the projection matrix of $\mathcal{R}^m$ onto the column domain $\mathcal{U}$ and the projection matrix of $\mathcal{R}^n$ onto the row domain $\mathcal{V}$ are, respectively, given by

$$P_{\mathcal{U}} = \sum_{i=1}^{r} u_i u_i^\top, \qquad P_{\mathcal{V}} = \sum_{i=1}^{r} v_i v_i^\top. \tag{4.8}$$

Since each u_i, $i = 1, ..., r$, is $u_i \in \mathcal{U}$, we have $P_{\mathcal{U}} u_i = u_i$. Hence, application of $P_{\mathcal{U}}$ to Eq. (4.4) from left does not cause any change. Similarly, we have $P_{\mathcal{V}} v_i = v_i$ for the rows. Hence, application of $P_{\mathcal{V}}$ to Eq. (4.4) from right does not cause any change. It follows that the following equalities hold:

$$P_{\mathcal{U}} A = A, \qquad A P_{\mathcal{V}} = A. \tag{4.9}$$

4.4 MATRIX REPRESENTATION

As we did for Eq. (4.4), we can rewrite Eq. (3.7) in the form

$$\begin{aligned} A &= \begin{pmatrix} \sigma_1 u_1 & \cdots & \sigma_r u_r \end{pmatrix} \begin{pmatrix} v_1^\top \\ \vdots \\ v_r^\top \end{pmatrix} = \begin{pmatrix} u_1 & \cdots & u_r \end{pmatrix} \begin{pmatrix} \sigma_1 & & \\ & \ddots & \\ & & \sigma_r \end{pmatrix} \begin{pmatrix} v_1^\top \\ \vdots \\ v_r^\top \end{pmatrix} \\ &= U \begin{pmatrix} \sigma_1 & & \\ & \ddots & \\ & & \sigma_r \end{pmatrix} V^\top, \qquad \sigma_1 \geq \cdots \geq \sigma_r > 0, \end{aligned} \tag{4.10}$$

where

$$U = \begin{pmatrix} u_1 & \cdots & u_r \end{pmatrix}, \qquad V = \begin{pmatrix} v_1 & \cdots & v_r \end{pmatrix} \tag{4.11}$$

are the $m \times r$ and $n \times r$ matrices consisting of singular vectors $\boldsymbol{u}_1, \ldots, \boldsymbol{u}_r$ and $\boldsymbol{v}_1, \ldots, \boldsymbol{v}_r$ as columns, respectively. Rewriting Eq. (4.5) in the same way results in the transpose of Eq. (4.10) on both sides.

Since the r columns of the matrices $\boldsymbol{U}$ and $\boldsymbol{V}$ are orthonormal systems, we obtain the identities ($\hookrightarrow$ Problem 4.4)

$$\boldsymbol{U}^\top \boldsymbol{U} = \boldsymbol{I}, \qquad \boldsymbol{V}^\top \boldsymbol{V} = \boldsymbol{I}, \tag{4.12}$$

where the right sides are the $r \times r$ identity matrix. We also obtain the following identities ($\hookrightarrow$ Problem 4.5):

$$\boldsymbol{U}\boldsymbol{U}^\top = \boldsymbol{P}_{\mathcal{U}}, \qquad \boldsymbol{V}\boldsymbol{V}^\top = \boldsymbol{P}_{\mathcal{V}}. \tag{4.13}$$

As mentioned in Section 4.2, we can extend the r left singular vectors $\boldsymbol{u}_1, \ldots, \boldsymbol{u}_r$ to an orthonormal basis $\{\boldsymbol{u}_1, \ldots, \boldsymbol{u}_r, \boldsymbol{u}_{r+1}, \ldots, \boldsymbol{u}_m\}$ of $\mathcal{R}^m$ and r right singular vectors $\boldsymbol{v}_1, \ldots, \boldsymbol{v}_r$ to an orthonormal basis $\{\boldsymbol{v}_1, \ldots, \boldsymbol{v}_r, \boldsymbol{v}_{r+1}, \ldots, \boldsymbol{v}_n\}$ of $\mathcal{R}^n$. If we let

$$\boldsymbol{U}' = \begin{pmatrix} \boldsymbol{u}_1 & \cdots & \boldsymbol{u}_m \end{pmatrix}, \qquad \boldsymbol{V}' = \begin{pmatrix} \boldsymbol{v}_1 & \cdots & \boldsymbol{v}_n \end{pmatrix}, \tag{4.14}$$

they are $m \times m$ and $n \times n$ orthogonal matrices, respectively. Then, Eq (4.10) is rewritten as

$$\boldsymbol{A} = \boldsymbol{U}' \begin{pmatrix} \sigma_1 & & & \\ & \ddots & & \\ & & \sigma_r & \\ & & & \\ & & & \end{pmatrix} \boldsymbol{V}'^\top, \tag{4.15}$$

where the middle matrix on the right side is an $m \times n$ matrix with $\sigma_1, \ldots, \sigma_r$ as the diagonal elements starting from the top left corner, all other elements being 0. Many numerical software tools for singular value decomposition output the result in this form.

4.5 GLOSSARY AND SUMMARY

column domain: The rD subspace of $\mathcal{R}^n$ spanned by the columns of an $m \times n$ matrix $\boldsymbol{A}$, where r is the rank of $\boldsymbol{A}$. The rD subspace $\mathcal{V}$ spanned by its rows is its "row domain."

singular value: For a nonzero matrix $\boldsymbol{A}$ ($\neq \boldsymbol{O}$), a positive number σ such that $\boldsymbol{A}\boldsymbol{v} = \sigma\boldsymbol{u}$ and $\boldsymbol{A}^\top\boldsymbol{u} = \sigma\boldsymbol{v}$ hold is a "singular value" of $\boldsymbol{A}$; the vectors $\boldsymbol{u}$ and $\boldsymbol{v}$ are called, respectively, the "left singular vector" and the "right singular vector" (generically the "singular vectors") for σ.

singular value decomposition: The expression of an arbitrary matrix $\boldsymbol{A}$ ($\neq \boldsymbol{O}$) in terms of its singular values σ_i and singular vectors $\boldsymbol{u}_i$, $\boldsymbol{v}_i$, $i = 1, \ldots, r$, in the form $\boldsymbol{A} = \sum_{i=1}^r \sigma_i \boldsymbol{u}_i \boldsymbol{v}_i^\top$, where r is the rank of $\boldsymbol{A}$.

- An arbitrary matrix $\boldsymbol{A}$ ($\neq \boldsymbol{O}$) has r singular values σ_i (> 0) and the corresponding singular vectors $\boldsymbol{u}_i$ and $\boldsymbol{v}_i$, $i = 1, ..., r$, where r is the rank of $\boldsymbol{A}$.
- Software tools are available for computing singular values and singular vectors.
- Singular vectors $\{\boldsymbol{u}_i\}$ and $\{\boldsymbol{v}_i\}$, $i = 1, ..., r$, both form orthonormal systems.
- The singular values σ_i, $i = 1, ..., r$, of $\boldsymbol{A}$ equal the square roots of the eigenvalues of $\boldsymbol{A}\boldsymbol{A}^\top$ and $\boldsymbol{A}\boldsymbol{A}^\top$; the corresponding eigenvectors $\boldsymbol{u}_i$ and $\boldsymbol{v}_i$, $i = 1, ..., r$, equal the singular vectors of $\boldsymbol{A}$.
- An arbitrary matrix $\boldsymbol{A}$ ($\neq \boldsymbol{O}$) is expressed in terms of its singular values σ_i and singular vectors $\boldsymbol{u}_i$ and $\boldsymbol{v}$, $i = 1, ..., r$, in the form $\boldsymbol{A} = \sum_{i=1}^{r} \sigma_i \boldsymbol{u}_i \boldsymbol{v}_i^\top$, called "singular value decomposition."
- The singular value decomposition of $\boldsymbol{A}$ is written as $\boldsymbol{A} = \boldsymbol{U}\boldsymbol{\Sigma}\boldsymbol{V}^\top$, where $\boldsymbol{U}$ and $\boldsymbol{V}$ are the matrices consisting of the left and right singular vectors of $\boldsymbol{A}$ as their columns, respectively, and $\boldsymbol{\Sigma}$ is a diagonal matrix with the singular values as its diagonal elements.
- The rank r (= the number of linearly independent columns and rows) of $\boldsymbol{A}$ equals the number of its singular values, and their singular vectors $\{\boldsymbol{u}_i\}$ and $\{\boldsymbol{v}_i\}$, $i = 1, ..., r$, define orthonormal bases of the column domain $\mathcal{U}$ and the row domain $\mathcal{V}$ of $\boldsymbol{A}$, respectively.
- For the singular vectors $\{\boldsymbol{u}_i\}$ and $\{\boldsymbol{v}_i\}$, $i = 1, ..., r$, of $\boldsymbol{A}$, the matrices $\boldsymbol{P}_{\mathcal{U}} = \sum_{i=1}^{r} \boldsymbol{u}_i \boldsymbol{u}_i^\top$ and $\boldsymbol{P}_{\mathcal{V}} = \sum_{i=1}^{r} \boldsymbol{v}_i \boldsymbol{v}_i^\top$ are the projection matrices onto the column domain $\mathcal{U}$ and the row domain $\mathcal{V}$ of $\boldsymbol{A}$, respectively.

4.6 SUPPLEMENTAL NOTES

For singular value decomposition, the term "decomposition" has been understood in two different senses. One interpretation is, as in the case of spectral decomposition, that we decompose a matrix into the "sum" of singular value components in the form of Eq. (4.4). The other interpretation is, as in Eq. (3.7), that we decompose a matrix into the "product" of three matrices in the form of Eq. (4.10), i.e., into the two matrices having orthonormal columns and a diagonal matrix consisting of singular values.

While the former interpretation is a natural extension of the spectral decomposition, the latter can be given a geometric interpretation. Consider the case of $m = n = r$:

$$\boldsymbol{A} = \boldsymbol{U} \begin{pmatrix} \sigma_1 & & \\ & \ddots & \\ & & \sigma_n \end{pmatrix} \boldsymbol{V}^\top. \tag{4.16}$$

Here, $\boldsymbol{U}$ and $\boldsymbol{V}$ are both $n \times n$ orthogonal matrices. This equation implies that the transformation of $\mathcal{R}^n$ represented by the matrix $\boldsymbol{A}$ is a composition of a rotation $\boldsymbol{V}^\top$ followed by an compression/expansion along each coordinate axes, which is called *general shear* in material mechanics, and a final rotation $\boldsymbol{U}$. The general shear matrix is further decomposed into

$$\sigma \begin{pmatrix} 1 & & \\ & \ddots & \\ & & 1 \end{pmatrix} + \begin{pmatrix} \sigma_1 - \sigma & & \\ & \ddots & \\ & & \sigma_n - \sigma \end{pmatrix}, \qquad \sigma = \sigma_1 + \cdots + \sigma_r, \tag{4.17}$$

The first matrix is a scalar multiple of the identity matrix $\boldsymbol{I}$ and the second is a traceless diagonal matrix (its trace is 0). In material mechanics, the first matrix corresponds to *hydrostatic compression* and the second matrix *pure shear*. Thus, in term of material mechanics, deformation of material is decomposed into rotations, hydrostatic compression, and pure shear, each having an invariant meaning. See Kanatani [7] for the group-theoretical interpretation.

4.7 PROBLEMS

4.1. Show that for any matrix $\boldsymbol{A}$, the matrices $\boldsymbol{A}\boldsymbol{A}^\top$ and $\boldsymbol{A}^\top\boldsymbol{A}$ are both positive semidefinite symmetric matrices, i.e., symmetric matrices whose eigenvalues are all positive or zero.

4.2.* Suppose one of the two matrices $\boldsymbol{A}\boldsymbol{A}^\top$ and $\boldsymbol{A}^\top\boldsymbol{A}$ has a positive eigenvalue σ for $\boldsymbol{A} \neq \boldsymbol{O}$. Show that it is also the eigenvalue of the other matrix and that their eigenvectors $\boldsymbol{u}$ and $\boldsymbol{v}$ are related by Eq. (4.1).

4.3. Show the following:

(1) If $\boldsymbol{A}\boldsymbol{A}^\top\boldsymbol{u} = \boldsymbol{0}$, then $\boldsymbol{A}^\top\boldsymbol{u} = \boldsymbol{0}$.

(2) If $\boldsymbol{A}^\top\boldsymbol{A}\boldsymbol{v} = \boldsymbol{0}$, then $\boldsymbol{A}\boldsymbol{v} = \boldsymbol{0}$.

4.4. Show that Eq. (4.12) holds.

4.5. Show that Eq. (4.13) holds.

CHAPTER 5
Pseudoinverse

A square matrix has its inverse if it is nonsingular. In this chapter, we define the "pseudoinverse" that extends the inverse to an arbitrary rectangular matrix. While the usual inverse is defined in such a way that its product with the original matrix equals the identity, the product of the pseudoinverse with the original matrix is not the identity but the projection matrix onto the space spanned by its columns and rows. Since all the columns and rows of a nonsingular matrix are linearly independent, they span the entire space, and the identity is the projection matrix onto it. In this sense, the pseudoinverse is a natural extension to the usual inverse. Next, we show that vectors, i.e., $n \times 1$ or $1 \times n$ matrices, also have their pseudoinverses. We point out that we need a special care for computing the pseudoinverse of a matrix whose elements are obtained by measurement in the presence of noise. We also point out that the error in such matrices is evaluated in the "matrix norm."

5.1 PSEUDOINVERSE

If an $m \times n$ matrix $\boldsymbol{A}$ ($\neq \boldsymbol{O}$) has the singular value decomposition in the form of Eq. (4.4), its *pseudoinverse*, or *generalized inverse*, of the *Moore–Penrose type* is defined to be the following $n \times m$ matrix:

$$\boldsymbol{A}^{-} = \frac{\boldsymbol{v}_1 \boldsymbol{u}_1^\top}{\sigma_1} + \cdots + \frac{\boldsymbol{v}_r \boldsymbol{u}_r^\top}{\sigma_r}. \tag{5.1}$$

If $\boldsymbol{A}$ is a nonsingular matrix, this coincides with the inverse $\boldsymbol{A}^{-1}$ of $\boldsymbol{A}$ ($\hookrightarrow$ Problem 5.1). In this sense, the pseudoinverse is a generalization of the inverse.

We will give a brief overview of pseudoinverses that are not of Moore–Penrose type in Section 5.6. Some authors write $\boldsymbol{A}^{-}$ for a "general" pseudoinverse and specifically write $\boldsymbol{A}^{+}$ for that of the Moore–Penrose type to make a distinction. However, we mainly focus only on the Moore–Penrose type in subsequent chapters.

If we define the matrices $\boldsymbol{U}$ and $\boldsymbol{V}$ as in Eq. (4.11), Eq. (4.4) can be written, in the same way as Eq. (4.10), in the matrix form

$$\boldsymbol{A}^{-} = \boldsymbol{V} \begin{pmatrix} 1/\sigma_1 & & \\ & \ddots & \\ & & 1/\sigma_r \end{pmatrix} \boldsymbol{U}^\top. \tag{5.2}$$

5.2 PROJECTION ONTO THE COLUMN AND ROW DOMAINS

While the inverse of a nonsingular matrix is defined so that the product is the identity matrix, the product of the pseudoinverse and the original matrix is note necessarily the identity. In fact, noting that $\{\boldsymbol{u}_i\}$ and $\{\boldsymbol{v}_i\}$, $i = 1, ..., r$, are orthonormal systems, we obtain from Eqs. (4.4) and (5.1) the following relationships (↪ Eq. (4.8)):

$$\begin{aligned}
\boldsymbol{A}\boldsymbol{A}^- &= \Big(\sum_{i=1}^{r} \sigma_i \boldsymbol{u}_i \boldsymbol{v}_i^\top\Big)\Big(\sum_{j=1}^{r} \frac{\boldsymbol{v}_j \boldsymbol{u}_j^\top}{\sigma_j}\Big) = \sum_{i,j=1}^{r} \frac{\sigma_i}{\sigma_j} \boldsymbol{u}_i (\boldsymbol{v}_i^\top \boldsymbol{v}_j) \boldsymbol{u}_j^\top = \sum_{i,j=1}^{r} \frac{\sigma_i}{\sigma_j} \langle \boldsymbol{v}_i, \boldsymbol{u}_j \rangle \boldsymbol{u}_i \boldsymbol{u}_j^\top \\
&= \sum_{i,j=1}^{r} \frac{\sigma_i}{\sigma_j} \delta_{ij} \boldsymbol{u}_i \boldsymbol{u}_j^\top = \sum_{i=1}^{r} \boldsymbol{u}_i \boldsymbol{u}_i^\top = \boldsymbol{P}_{\mathcal{U}},
\end{aligned} \tag{5.3}$$

$$\begin{aligned}
\boldsymbol{A}^-\boldsymbol{A} &= \Big(\sum_{i=1}^{r} \frac{\boldsymbol{v}_i \boldsymbol{u}_i^\top}{\sigma_i}\Big)\Big(\sum_{j=1}^{r} \sigma_j \boldsymbol{u}_j \boldsymbol{v}_j^\top\Big) = \sum_{i,j=1}^{r} \frac{\sigma_i}{\sigma_j} \boldsymbol{v}_i (\boldsymbol{u}_i^\top \boldsymbol{u}_j) \boldsymbol{v}_j^\top = \sum_{i,j=1}^{r} \frac{\sigma_i}{\sigma_j} \langle \boldsymbol{u}_i, \boldsymbol{u}_j \rangle \boldsymbol{v}_i \boldsymbol{v}_j^\top \\
&= \sum_{i,j=1}^{r} \frac{\sigma_i}{\sigma_j} \delta_{ij} \boldsymbol{v}_i \boldsymbol{v}_j^\top = \sum_{i=1}^{r} \boldsymbol{v}_i \boldsymbol{v}_i^\top = \boldsymbol{P}_{\mathcal{V}}.
\end{aligned} \tag{5.4}$$

Note that when the Kronecker delta δ_{ij} appears in summations $\sum$ over i or j (or both), only terms for which $i = j$ survive. From the above equations, we find that *the products $\boldsymbol{A}\boldsymbol{A}^-$ and $\boldsymbol{A}^-\boldsymbol{A}$ are the projection matrices onto the column domain $\mathcal{U}$ and the row domain $\mathcal{V}$, respectively* (↪ Problem 5.2), i.e.,

$$\boldsymbol{A}\boldsymbol{A}^- = \boldsymbol{P}_{\mathcal{U}}, \qquad \boldsymbol{A}^-\boldsymbol{A} = \boldsymbol{P}_{\mathcal{V}}. \tag{5.5}$$

For a nonsingular matrix, its columns and rows are linearly independent, spanning the entire space, and the projection matrix onto the entire space is the identity matrix (↪ Eq. (2.11)). Hence, the pseudoinverse is a natural extension of the inverse.

Since $\boldsymbol{P}_{\mathcal{U}}\boldsymbol{x} = \boldsymbol{x}$ for any $\boldsymbol{x} \in \mathcal{U}$, the matrix $\boldsymbol{P}_{\mathcal{U}}$ plays the role of the identity in the column domain $\mathcal{U}$. Hence, the first equation of Eq. (5.5) states that *$\boldsymbol{A}^-$ defines the inverse transformation of $\boldsymbol{A}$ in the column domain $\mathcal{U}$*. Similarly, since $\boldsymbol{P}_{\mathcal{V}}\boldsymbol{x} = \boldsymbol{x}$ for any $\boldsymbol{x} \in \mathcal{V}$, the matrix $\boldsymbol{P}_{\mathcal{V}}$ plays the role of the identity in the row domain $\mathcal{V}$. Hence, the second equation of Eq. (5.5) states that *$\boldsymbol{A}^-$ defines the inverse transformation of $\boldsymbol{A}$ in the row domain $\mathcal{V}$*.

Since $\boldsymbol{P}_{\mathcal{U}}\boldsymbol{u}_i = \boldsymbol{u}_i$ and $\boldsymbol{P}_{\mathcal{V}}\boldsymbol{v}_i = \boldsymbol{v}_i$ from the definition of the projection matrices $\boldsymbol{P}_{\mathcal{U}}$ and $\boldsymbol{P}_{\mathcal{V}}$, we obtain the identities

$$\boldsymbol{P}_{\mathcal{V}}\boldsymbol{A}^- = \boldsymbol{A}^-, \qquad \boldsymbol{A}^-\boldsymbol{P}_{\mathcal{U}} = \boldsymbol{A}^- \tag{5.6}$$

for the pseudoinverse $\boldsymbol{A}^-$ of Eq. (5.1) in the same way as Eq. (4.9). From this, we obtain the fundamental identities

$$\boldsymbol{A}^-\boldsymbol{A}\boldsymbol{A}^- = \boldsymbol{A}^-, \tag{5.7}$$

$$AA^{-}A = A \tag{5.8}$$

for the pseudoinvers. Equation (5.7) is obtained by combining Eq. (5.4) and the first equation of Eq. (5.6). Alternatively, we may combine Eq. (5.3) and the second equation of Eq. (5.6). Equation (5.8) is, on the other hand, obtained by combining Eq. (5.3) and the first equation of Eq. (4.9). Alternatively, we may combine Eq. (5.4) and the second equation of Eq. (4.9). Equations (5.7) and (5.8) can also be obtained from the matrix representation of Eq. (5.2) ($\hookrightarrow$ Problem 5.3).

5.3 PSEUDOINVERSE OF VECTORS

An nD vector $\boldsymbol{a}$ is an $n \times 1$ matrix, so it has its pseudoinverse. If $\boldsymbol{a} \neq \boldsymbol{0}$, its singular value decomposition is given by

$$\boldsymbol{a} = \|\boldsymbol{a}\|\Big(\frac{\boldsymbol{a}}{\|\boldsymbol{a}\|}\Big) \cdot 1. \tag{5.9}$$

The column domain is the 1D space spanned by the unit vector $\boldsymbol{u} = \boldsymbol{a}/\|\boldsymbol{a}\|$, and the row domain is $\mathcal{R}^1$ (= the set of real numbers) whose basis is 1. The singular value is $\|\boldsymbol{a}\|$. Hence, the pseudoinverse $\boldsymbol{a}^-$ is given by

$$\boldsymbol{a}^{-} = \frac{1}{\|\boldsymbol{a}\|} 1 \cdot \Big(\frac{\boldsymbol{a}}{\|\boldsymbol{a}\|}\Big)^{\top} = \frac{\boldsymbol{a}^{\top}}{\|\boldsymbol{a}\|^2}. \tag{5.10}$$

In other words, *the transposed row vector divided by its square length* $\|\boldsymbol{a}\|^2$.

From Eqs. (4.4), (4.5), and (5.1), we see that $(\boldsymbol{A}^{\top})^{-} = (\boldsymbol{A}^{-})^{\top}$, which we simply write $\boldsymbol{A}^{-\top}$. Hence, the pseudoinverse of a row vector $\boldsymbol{a}^{\top}$ regarded as a 1×3 matrix is given by

$$\boldsymbol{a}^{-\top} = \frac{\boldsymbol{a}}{\|\boldsymbol{a}\|^2}. \tag{5.11}$$

If we write the unit direction vector along vector $\boldsymbol{a}$ as $\boldsymbol{u} = \boldsymbol{a}/\|\boldsymbol{a}\|$, the product of the pseudoinverse $\boldsymbol{a}^-$ and the vector $\boldsymbol{a}$ is

$$\boldsymbol{a}\boldsymbol{a}^{-} = \boldsymbol{a}\frac{\boldsymbol{a}^{\top}}{\|\boldsymbol{a}\|^2} = \boldsymbol{u}\boldsymbol{u}^{\top}, \tag{5.12}$$

which is the projection matrix onto the direction of the vector $\boldsymbol{u}$. On the other hand, we see that

$$\boldsymbol{a}^{-}\boldsymbol{a} = \frac{\boldsymbol{a}^{\top}}{\|\boldsymbol{a}\|^2}\boldsymbol{a} = \frac{\langle \boldsymbol{a}, \boldsymbol{a} \rangle}{\|\boldsymbol{a}\|^2} = 1. \tag{5.13}$$

Note that 1 (= the 1×1 identity matrix) is the projection matrix onto $\mathcal{R}^1$.

5.4 RANK-CONSTRAINED PSEUDOINVERSE

In Chapters 3 and 4, we pointed out that various software tools are available for computing eigenvalues, eigenvectors, and the singular value decomposition. However, this is not so for pseudoinverses. Basically, there is no software tool to automatically compute the pseudoinverse; if such a tool is offered, we should not use it. This is because almost all computations that arise in physics and engineering are based of observation data obtained by measurement devices and sensors. Hence, all data contain noise to some extent. As a result, if we compute the singular value decomposition by Eq. (4.4), all the singular values σ_i are generally positive. If some of σ_i are ideally 0 but are computed to be nonzero due to the noise, we may obtain unrealistic values using Eq. (5.1) due to the term $1/\sigma_i$.

Of course, this is not limited to pseudoinverses; it also applies to the computation of the usual inverses, e.g., when we use Eq. (3.11) to a matrix which is not really nonsingular. Note that while the inverse is defined only for nonsingular matrices, the pseudoinverse is defined for all nonzero matrices. The important thing is that *we need to know the rank* before computing the pseudoinverse, using Eq. (5.1).

A simple way to judge the rank r of an $m \times n$ matrix $\boldsymbol{A}$ obtained from measurement data is to compute the singular value decomposition

$$\boldsymbol{A} = \sigma_1 \boldsymbol{u}_1 \boldsymbol{v}_1^\top + \cdots + \sigma_l \boldsymbol{u}_l \boldsymbol{v}_l^\top, \qquad \sigma_1 \geq \cdots \geq \sigma_l, \tag{5.14}$$

by letting $l = \min(m, n)$, and to find a value r such that

$$\sigma_{r+1} \approx 0, \quad \ldots, \quad \sigma_l \approx 0 \tag{5.15}$$

for the trailing singular values. We regard them as noise and retain the singular values up to σ_r, i.e., we replace $\boldsymbol{A}$ by

$$(\boldsymbol{A})_r = \sigma_1 \boldsymbol{u}_1 \boldsymbol{v}_1^\top + \cdots + \sigma_r \boldsymbol{u}_r \boldsymbol{v}_r^\top, \tag{5.16}$$

truncating the trailing singular values, and compute its pseudoinverse $(\boldsymbol{A})_r^-$, which we call the *rank-constrained pseudoinverse* (or *generalized inverse*). But how should we find the threshold for truncating the small singular values?

For a mathematical computation where the data are exact, possible errors are due to the rounding of finite length numerical computation. Hence, we can use the smallest number that can be digitally represented in a computer, which is called the *machine epsilon*, as the threshold; some software tools are so designed. For computations appearing in physics and engineering that involve observation data, however, it is generally difficult to estimate the error magnitude; it is different from problem to problem.

On the other hand, most application problems of physics and engineering are derived from some fundamental principles or laws. Hence, it is usually possible to do theoretical analysis based on such principles or laws to predict the rank r in an ideal case where the measurement devices or sensors are assumed to be noiseless. Then, we can use the theoretical rank r regardless of the magnitude of the singular values to be truncated and compute the rank-constrained pseudoinverse of Eq. (5.16).

We are then interested in estimating the difference between the rank-constrained matrix obtained by truncating the trailing singular values and the original matrix. This is done using the matrix norm.

5.5 EVALUATION BY MATRIX NORM

The *matrix norm* of an $m \times n$ matrix $\boldsymbol{A} = \left(A_{ij}\right)$ is defined by

$$\|\boldsymbol{A}\| = \sqrt{\sum_{i=1}^{m}\sum_{j=1}^{n} A_{ij}^2}. \tag{5.17}$$

This is called the *Frobenius norm* or the *Euclid norm*, and the following holds (↪ Problems 5.4 and 5.5):

$$\|\boldsymbol{A}\|^2 = \mathrm{tr}(\boldsymbol{A}^\top \boldsymbol{A}) = \mathrm{tr}(\boldsymbol{A}\boldsymbol{A}^\top). \tag{5.18}$$

Here, tr denotes the matrix trace.

Using Eq. (5.18), we can evaluate the difference between the matrix $\boldsymbol{A}$ of Eq. (5.14) and the matrix $(\boldsymbol{A})_r$ of Eq. (5.16), measured in the square matrix norm, as follows:

$$\begin{aligned}
\|\boldsymbol{A} - (\boldsymbol{A})_r\|^2 &= \|\sum_{i=r+1}^{l} \sigma_i \boldsymbol{u}_i \boldsymbol{v}_i^\top\|^2 = \mathrm{tr}\Big(\Big(\sum_{i=r+1}^{l} \sigma_i \boldsymbol{u}_i \boldsymbol{v}_i^\top\Big)^\top \sum_{j=r+1}^{l} \sigma_j \boldsymbol{u}_j \boldsymbol{v}_j^\top\Big) \\
&= \mathrm{tr}\Big(\sum_{i,j=r+1}^{l} \sigma_i\sigma_j \boldsymbol{v}_i \boldsymbol{u}_i^\top \boldsymbol{u}_j \boldsymbol{v}_j^\top\Big) = \mathrm{tr}\Big(\sum_{i,j=r+1}^{l} \sigma_i\sigma_j \boldsymbol{v}_i \langle \boldsymbol{u}_i, \boldsymbol{u}_j\rangle \boldsymbol{v}_j^\top\Big) \\
&= \mathrm{tr}\Big(\sum_{i,j=r+1}^{l} \delta_{ij}\sigma_i\sigma_j \boldsymbol{v}_i \boldsymbol{v}_j^\top\Big) = \sum_{i=r+1}^{l} \sigma_i^2 \mathrm{tr}(\boldsymbol{v}_i \boldsymbol{v}_i^\top) = \sum_{i=r+1}^{l} \sigma_i^2.
\end{aligned} \tag{5.19}$$

Note that $\mathrm{tr}(\boldsymbol{v}_i \boldsymbol{v}_i^\top) = \|\boldsymbol{v}_i\|^2 = 1$ holds from Eq. (2.28). Thus, we conclude that the difference between the matrix $\boldsymbol{A}$ and the matrix $(\boldsymbol{A})_r$ obtained by truncating singular values is equal to *the square sum of the truncated singular values*[1]:

$$\|\boldsymbol{A} - (\boldsymbol{A})_r\| = \sqrt{\sigma_{r+1}^2 + \cdots + \sigma_l^2}. \tag{5.20}$$

This can be also derived using the matrix representation (↪ Problem 5.6).

[1]It can be shown that for a given matrix $\boldsymbol{A}$, the matrix $\boldsymbol{A}'$ of the same size that minimizes $\|\boldsymbol{A} - \boldsymbol{A}'\|$ subject to the constraint $\mathrm{rank}(\boldsymbol{A}') = r$ is given by $\boldsymbol{A}' = (\boldsymbol{A})_r$. The proof is rather complicated [1].

5.6 GLOSSARY AND SUMMARY

machine epsilon: The smallest number that can be digitally represented in a computer.

matrix norm: The number $\|A\| = \sqrt{\sum_{i=1}^{m}\sum_{j=1}^{n} A_{ij}^2}$ that measures the magnitude of an $m \times n$ matrix $A = \left(A_{ij}\right)$, also called the "Frobenius norm" or the "Euclid norm."

pseudoinverse (generalized inverse): For a matrix with singular value decomposition $A = \sum_{i=1}^{r} \sigma_i u_i v_i^\top$, the matrix $A^- = \sum_{i=1}^{r}(1/\sigma_i) v_i u_i^\top$ is its "pseudoinverse" or "generalized inverse" (of the "Moore–Penrose type" to be precise).

rank-constrained pseudoinverse (generalized inverse): For a matrix with singular value decomposition $A = \sum_{i=1}^{r} \sigma_i u_i v_i^\top$, the pseudoinverse $(A)_r^-$ of the matrix $(A)_r = \sum_{i=1}^{r} \sigma_i u_i v_i^\top$ obtained by replacing the smaller singular values σ_{r+1}, ..., σ_l by 0 is the "rank-constrained pseudoinverse (or generalized inverse)" of A to rank r.

- The pseudoinverse of a matrix A is obtained by replacing its singular values by their reciprocals in the singular value decomposition and by transposing the entire expression (Eq. (5.1)).
- An arbitrary matrix A ($\neq O$) has its pseudoinverse A^-.
- The matrices AA^- and A^-A are the projection matrices onto the column domain $\mathcal{U}$ and the row domain $\mathcal{V}$ of A, respectively (Eq. (5.5)).
- If the Kronecker delta δ_{ij} appears in a sum $\sum$ over i or j (or both), only terms for $i = j$ survive.
- The pseudoinverse A^- represents the inverse operation of A within the column domain $\mathcal{U}$ and the row domain $\mathcal{V}$ (Eq. (5.5)).
- The pseudoinverse of a vector is its transpose divided by its squared norm (Eqs. (5.10) and (5.11)).
- For computing pseudoinverse, we need to know the rank of the matrix.
- For a matrix whose elements are obtained by measurement, we infer its rank in the absence of noise by a theoretical consideration and compute the rank-constrained pseudoinverse.
- If smaller singular values are truncated and replaced by 0, the resulting deviation of the matrix, measured in matrix norm, equals the square root of the square sum of the truncated singular values (Eq. (5.20)).

5.7 SUPPLEMENTAL NOTES

Mathematically, Eq. (5.8) is the definition of the most *general* "pseudoinverse" $\boldsymbol{A}^-$. By adding various conditions, various other types of pseudoinverse are defined. If Eq. (5.7) is satisfied, $\boldsymbol{A}^-$ is said to be a *reflexive pseudoinverse*. If $\boldsymbol{A}\boldsymbol{A}^-$ and $\boldsymbol{A}^-\boldsymbol{A}$ are both symmetric matrices, it is of the Moor–Penrose type.

Let us briefly overview the background and motivations behind these. An $m \times n$ matrix $\boldsymbol{A}$ defines a linear mapping from $\mathcal{R}^n$ to $\mathcal{R}^m$. If we let $\boldsymbol{A} = \sum_{k=1}^{r} \sigma_i \boldsymbol{u}_i \boldsymbol{v}_i^\top$ be the singular decomposition of $\boldsymbol{A}$, it maps $\mathcal{R}^n$ onto the rD subset $\mathcal{U}$ of $\mathcal{R}^m$ spanned by $\{\boldsymbol{u}_1, ..., \boldsymbol{u}_r\}$, which we called the "column domain" of $\boldsymbol{A}$; this is also denoted by $\mathrm{Im}(\boldsymbol{A})$ and called the *image* of $\boldsymbol{A}$. We can extend $\{\boldsymbol{u}_1, ..., \boldsymbol{u}_r\}$ to an orthonormal basis $\{\boldsymbol{u}_1, ..., \boldsymbol{u}_m\}$ of $\mathcal{R}^m$ by, say, Schmidt orthogonalization. The $(m-r)$D subspace spanned by $\{\boldsymbol{u}_{r+1}, ..., \boldsymbol{u}_m\}$ is the orthogonal complement $\mathcal{U}^\perp$ of $\mathcal{U}$, and we obtain

$$\mathcal{R}^m = \mathcal{U} \oplus \mathcal{U}^\perp (= \mathrm{Im}(\boldsymbol{A}) \oplus \mathcal{U}^\perp), \tag{5.21}$$

which is the direct sum decomposition of $\mathcal{R}^m$.

On the other hand, the rD subset $\mathcal{V}$ of $\mathcal{R}^n$ spanned by $\{\boldsymbol{v}_1, ..., \boldsymbol{v}_r\}$, which we called the "row domain" and denoted by $\mathcal{V}$, is mapped by $\boldsymbol{A}$ to $\mathcal{U}$ one-to-one via $\boldsymbol{v}_i \mapsto \sigma_i \boldsymbol{u}_i$. If we extend $\{\boldsymbol{v}_1, ..., \boldsymbol{v}_r\}$ to an orthonormal basis $\{\boldsymbol{v}_1, ..., \boldsymbol{v}_n\}$ of $\mathcal{R}^n$, the $(n-r)$D subspace spanned by $\{\boldsymbol{v}_{r+1}, ..., \boldsymbol{u}_n\}$ is the orthogonal complement $\mathcal{V}^\perp$ of $\mathcal{V}$ and is mapped by $\boldsymbol{A}$ to $\boldsymbol{0} \in \mathcal{R}^m$; this is also denoted by $\ker(\boldsymbol{A})$ and called the *kernel* of $\boldsymbol{A}$ ($\hookrightarrow$ Section 2.6). We see that

$$\mathcal{R}^n = \mathcal{V} \oplus \mathcal{V}^\perp (= \mathcal{V} \oplus \ker(\boldsymbol{A})) \tag{5.22}$$

is the direct sum decomposition of $\mathcal{R}^n$.

Now, we want to define a mapping $\boldsymbol{A}^-$ from $\mathcal{R}^m$ to $\mathcal{R}^n$ that behaves like an inverse. If we map $\mathcal{R}^n$ to $\mathcal{R}^m$ by $\boldsymbol{A}$ and then map $\mathcal{R}^m$ back to $\mathcal{R}^n$ by $\boldsymbol{A}^-$, the composition $\boldsymbol{A}^-\boldsymbol{A}$ cannot be the identity (unless $n = m$ and $\boldsymbol{A}$ is nonsingular), but we require that applying $\boldsymbol{A}$ again will result in the same mapping as the original $\boldsymbol{A}$, i.e., $\boldsymbol{A}\boldsymbol{A}^{-1}\boldsymbol{A} = \boldsymbol{A}$. Such a mapping $\boldsymbol{A}^-$ (not unique) is called a "pseudoinverse" of $\boldsymbol{A}$.

The motivation behind such a mapping is this. For a given $\boldsymbol{b} \in \mathcal{R}^m$, the linear equation $\boldsymbol{A}\boldsymbol{x} = \boldsymbol{b}$ does not necessarily have a solution $\boldsymbol{x}$, but suppose there exists a particular solution $\boldsymbol{x}^*$ (not unique in general) such that $\boldsymbol{A}\boldsymbol{x}^* = \boldsymbol{b}$. If a pseudoinverse $\boldsymbol{A}^-$ such that $\boldsymbol{A}\boldsymbol{A}^-\boldsymbol{A} = \boldsymbol{A}$ exists, we have $\boldsymbol{A}\boldsymbol{A}^-\boldsymbol{A}\boldsymbol{x}^* = \boldsymbol{A}\boldsymbol{x}^*$, i.e., $\boldsymbol{A}\boldsymbol{A}^-\boldsymbol{b} = \boldsymbol{b}$, or $\boldsymbol{A}(\boldsymbol{A}^-\boldsymbol{b}) = \boldsymbol{b}$. This means that $\boldsymbol{x} = \boldsymbol{A}^-\boldsymbol{b}$ gives a solution of $\boldsymbol{A}\boldsymbol{x} = \boldsymbol{b}$ for any pseudoinverse $\boldsymbol{A}^-$.

Such a pseudoinverse $\boldsymbol{A}^-$ can be defined as follows. Since the mapping $\boldsymbol{A}$ from $\mathcal{V}$ to $\mathcal{U}$ is one-to-one, we can see from Eqs. (5.24) and (5.25) that we can define $\boldsymbol{A}^- : \mathcal{R}^m \to \mathcal{R}^n$ by requiring that $\mathcal{U}$ be mapped to $\mathcal{V}$ via $\boldsymbol{u}_i \mapsto \boldsymbol{v}_i/\sigma_i$ and $\mathcal{U}^\perp$ be mapped to $\mathcal{V}^\perp$ by an arbitrary linear mapping (not unique). Then, $\boldsymbol{A}^-\boldsymbol{A}$ maps the entire $\mathcal{R}^n$ to $\mathcal{V} \subset \mathcal{R}^n$, i.e., it is a (not necessarily orthogonal) projection of $\mathcal{R}^n$ onto $\mathcal{V}$ ($\hookrightarrow$ Section 2.6).

We may require A^- to be "reflexive" in the sense that if A^- is an pseudoinverse of A, then A is also a pseudoinverse of A^-, i.e., $A^-AA^- = A^-$. Then, AA^- is a (not necessarily orthogonal) projection of $\mathcal{R}^m$ onto $\mathcal{U}$.

In particular, if we define the mapping from $\mathcal{U}^\perp$ to $\mathcal{V}^\perp$ by demanding $\mathcal{U}^\perp \to \{\mathbf{0}\} \in \mathcal{V}^\perp$, the resulting pseudoinvers A^- is of the Moore–Penrose type given by Eq. (5.1). Since A maps $\mathcal{V}^\perp$ to $\mathbf{0} \in \mathcal{U}^\perp$, this pseudoinverse A^- is reflexive, and A^-A and AA^- are both (orthogonal) projections onto $\mathcal{V}$ and $\mathcal{U}$, respectively. Note that A^-A and AA^- are orthogonal projections if and only if they are symmetric matrices (↪ Problems 2.4 and 2.7). We can alternatively define the Moore–Penrose type by requiring that A^- be reflexive and both A^-A and AA^- be symmetric.

5.8 PROBLEMS

5.1. Show that if A is nonsingular, i.e., $m = n$ and its eigenvalues are all nonzero, i.e., $r = n$, Eq. (5.1) defines the inverse A^{-1} of A.

5.2. Using Eqs. (4.10) and (5.2), show that Eq. (5.5) holds.

5.3. Using Eqs. (4.10) and (5.2), show that Eq. (5.7) holds.

5.4. Show that the identity

$$\mathrm{tr}(AB) = \mathrm{tr}(BA) \tag{5.23}$$

holds for the matrix trace, where the sizes of the matrices are such that the products can be defined.

5.5. Show that the identity

$$\|A\| = \|AU\| = \|VA\| = \|VAU\| \tag{5.24}$$

holds for orthogonal matrices U and V having sizes such that the products can be defined.

5.6. Show that if matrix A has a singular value decomposition in the form of Eq. (4.10), its norm is given by

$$\|A\| = \sqrt{\sigma_1^2 + \cdots + \sigma_r^2}, \tag{5.25}$$

and hence Eq. (5.20) is obtained.

CHAPTER 6

Least-Squares Solution of Linear Equations

The pseudoinverse introduced in the preceding chapter is closely related to the least-squares method for linear equations. In fact, the theory of pseudoinverse has been studied in relation to minimization of the sum of squares of linear equations. The least-squares method usually requires solving an equation, called the "normal equation," obtained by letting the derivative of the sum of squares be zero. In this chapter, we show how a general solution is obtained without using differentiation or normal equations. As illustrative examples, we show the case of multiple equations of one variable and the case of a single multivariate equation.

6.1 LINEAR EQUATIONS AND LEAST SQUARES

Consider m simultaneous linear equations of n variables $x_1, \ldots, x_n$:

$$\begin{aligned} a_{11}x_1 + \cdots + a_{1n}x_n &= b_1 \\ &\vdots \\ a_{m1}x_1 + \cdots + a_{mn}x_n &= b_m. \end{aligned} \tag{6.1}$$

Using vectors and matrices, we can write these as

$$\boldsymbol{A}\boldsymbol{x} = \boldsymbol{b}, \tag{6.2}$$

where the $m \times n$ matrix $\boldsymbol{A}$ the nD vector $\boldsymbol{x}$, and the mD vector $\boldsymbol{b}$ are, respectively, defined by

$$\boldsymbol{A} = \begin{pmatrix} a_{11} & \cdots & a_{1n} \\ \vdots & \ddots & \vdots \\ a_{m1} & \cdots & a_{mn} \end{pmatrix}, \qquad \boldsymbol{x} = \begin{pmatrix} x_1 \\ \vdots \\ x_n \end{pmatrix}, \qquad \boldsymbol{b} = \begin{pmatrix} b_1 \\ \vdots \\ b_m \end{pmatrix}. \tag{6.3}$$

Hereafter, we assume that $\boldsymbol{A} \neq \boldsymbol{O}$.

As is well known, Eq. (6.2) has a unique solution when $n = m$ and the determinant of $\boldsymbol{A}$ is nonzero, i.e., $\boldsymbol{A}$ is nonsingular. For numerically computing the solution, the *Gaussian elimination* and the mathematically equivalent *LU-decomposition* are well known [18]. However, problems with $n \neq m$ frequently occur in many physic and engineering applications that involve observation data.

Each equation of Eq. (6.1) is interpreted to be a measurement process for estimating the n parameters $x_1, ..., x_n$. Theoretically, we only need n measurements for determining n parameters. However, we often repeat the measurement m $(> n)$ times, considering that the observations may contain noise. In some cases, there are constraints on measurement so that only m $(< n)$ observations are possible. In the case of $n \neq m$, a practical approach is to estimating x_1, ..., x_n that satisfy all the equations of Eq. (6.1) "sufficiently well." A typical strategy for this is to minimize the sum of the squares of the differences between the right and left sides the individual equations. Namely, we compute x_1, ..., x_n that minimize

$$J = (a_{11}x_1 + \cdots + a_{1n}x_n - b_1)^2 + \cdots + (a_{m1}x_1 + \cdots + a_{mn}x_n - b_m)^2. \tag{6.4}$$

This approach is called the *least-squares method*. The value J is called the *residual sum of squares*, or the *residual* for short. If we write Eq. (6.1) in the form of Eq. (6.2) using vectors and matrices, the residual J of Eq. (6.4) is written as

$$J = \|\boldsymbol{A}\boldsymbol{x} - \boldsymbol{b}\|^2. \tag{6.5}$$

However, the value of $\boldsymbol{x}$ that minimizes J may not be determined uniquely if we cannot repeat the observation a sufficient number of times. In that case, we choose from among multiple possibilities the value $\boldsymbol{x}$ that minimizes $\|\boldsymbol{x}\|^2$. This reflects the fact that in many physics and engineering problems the value $\|\boldsymbol{x}\|^2$ represents some physical quantity that, like heat generation or necessary energy, should desirably be as small as possible. In view of this, we call the value $\boldsymbol{x}$ for which (i) the residual J is minimized and (ii) $\|\boldsymbol{x}\|^2$ is minimum the *least-squares solution*.

The least-squares method was introduced by the German mathematician Karl Gauss (1777–1855) for computing the motion of planets from telescope observation data.[1] He also introduced various numerical techniques for efficiently solving simultaneous linear equations and accurately evaluating integrals, establishing the foundations of today's numerical analysis. In order to justify his least-squares approach, he introduced a mathematical model of numerical noise contained in observation data. Asserting that such a noise distribution is the most "normal," he called it the *normal distribution* (we will discuss this in detail in Section 7.2), which plays the fundamental role in today's statistical analysis. Many, mostly physicists and engineers, call it the *Gaussian distribution* in honor of Gauss. He also made various great contributions in pure mathematics, including the fundamental theorem of algebra (the Gauss theorem). At the same time, he established many differential and integral formulas of electromagnetics and fluid dynamics, which are the basis of today's physics.

6.2 COMPUTING THE LEAST-SQUARES SOLUTION

In general, the solution that minimizes Eq. (6.5) is obtained by differentiating J with respect to $\boldsymbol{x}$, letting the result be $\boldsymbol{0}$, and solving the resulting equation $\nabla_{\boldsymbol{x}} J = \boldsymbol{0}$, which is called the

[1] Some people say that the theory of least squares was first published independently by the French mathematician Adrien–Marie Legendre (1752–1833) several years before Gauss. This is one of the most famous priority disputes in the history of statistics.

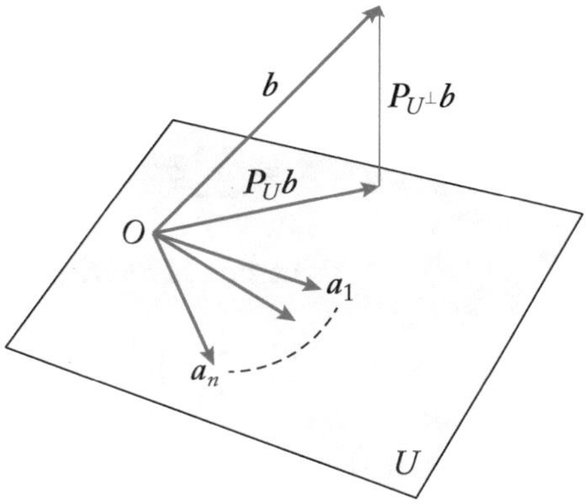

Figure 6.1: The vector $\boldsymbol{b}$ is projected onto the subspace $\mathcal{U}$ spanned by the columns $\boldsymbol{a}_1$, ..., $\boldsymbol{a}_n$ of $\boldsymbol{A}$.

normal equation. However, its solution has different forms, depending on whether $m > n$ or $m < n$ and whether $r = n$, $r = m$, or otherwise (↪ Problems 6.1–6.4). Here, we show that the most general form of the least-squares solution that encompasses all the cases is obtained, using the projection matrix and the pseudoinverse, without involving differentiation or solving the normal equation. This is based on the observation that projection minimizes the square norm, i.e., the sum of squares.

Let $\mathcal{U}$ be the column domain of $\boldsymbol{A}$ of Eq. (6.3), i.e., the space spanned by the columns of $\boldsymbol{A}$. This is a subspace of $\mathcal{R}^m$. From Eq. (2.12), we can express the square norm of a vector as the sum of the square norm of its projection onto the subspace $\mathcal{U}$ and the square norm of its rejection from it, i.e., the projection onto the orthogonal complement $\mathcal{U}^\perp$. Hence, the residual J of Eq. (6.5) can be written as

$$J = \|\boldsymbol{P}_\mathcal{U}(\boldsymbol{A}\boldsymbol{x} - \boldsymbol{b})\|^2 + \|\boldsymbol{P}_{\mathcal{U}^\perp}(\boldsymbol{A}\boldsymbol{x} - \boldsymbol{b})\|^2 = \|\boldsymbol{A}\boldsymbol{x} - \boldsymbol{P}_\mathcal{U}\boldsymbol{b}\|^2 + \|\boldsymbol{P}_{\mathcal{U}^\perp}\boldsymbol{b}\|^2. \tag{6.6}$$

Note that $\boldsymbol{P}_\mathcal{U}\boldsymbol{A}\boldsymbol{x} = \boldsymbol{A}\boldsymbol{x}$ and $\boldsymbol{P}_{\mathcal{U}^\perp}\boldsymbol{A}\boldsymbol{x} = \boldsymbol{0}$, because $\boldsymbol{A}\boldsymbol{x}$ is a linear combination of the columns of $\boldsymbol{A}$ and hence is included in the column domain $\mathcal{U}$. Since the last term of Eq. (6.6) does not contain $\boldsymbol{x}$, the least-squares solution should satisfy

$$\boldsymbol{A}\boldsymbol{x} = \boldsymbol{P}_\mathcal{U}\boldsymbol{b}, \qquad J = \|\boldsymbol{P}_{\mathcal{U}^\perp}\boldsymbol{b}\|^2. \tag{6.7}$$

This is interpreted as follows. Since $\boldsymbol{A}\boldsymbol{x} \in \mathcal{U}$, Eq. (6.2) evidently has no solution unless $\boldsymbol{b} \in \mathcal{U}$. Hence, *we replace the vector* $\boldsymbol{b}$ *by its projection* $\boldsymbol{P}_\mathcal{U}\boldsymbol{b}$ *onto* $\mathcal{U}$. Since the part of $\boldsymbol{b}$ that is outside the column domain $\mathcal{U}$ is $\boldsymbol{P}_{\mathcal{U}^\perp}\boldsymbol{b}$, the residual equals $\|\boldsymbol{P}_{\mathcal{U}^\perp}\boldsymbol{b}\|^2$ (Fig. 6.1).

We are assuming that $\boldsymbol{A} \neq \boldsymbol{O}$, so $\boldsymbol{A}$ has the singular value decomposition in the form of Eq. (4.4). The left side of the first equation of Eq. (6.7) is rewritten as

$$\boldsymbol{A}\boldsymbol{x} = \sum_{i=1}^{r} \sigma_i \boldsymbol{u}_i \boldsymbol{v}_i^\top \boldsymbol{x} = \sum_{i=1}^{r} \sigma_i \langle \boldsymbol{v}_i, \boldsymbol{x} \rangle \boldsymbol{u}_i. \tag{6.8}$$

From the expression of $\boldsymbol{P}_{\mathcal{U}}$ of Eq. (4.8), the right side is rewritten in the form

$$\boldsymbol{P}_{\mathcal{U}}\boldsymbol{b} = \sum_{i=1}^{r} \boldsymbol{u}_i \boldsymbol{u}_i^\top \boldsymbol{b} = \sum_{i=1}^{r} \langle \boldsymbol{u}_i, \boldsymbol{b} \rangle \boldsymbol{u}_i. \tag{6.9}$$

Since Eqs. (6.8) and (5.8) are expansions in terms of the orthonormal system $\{\boldsymbol{u}_i\}$ (↪ Appendix A.7), we have $\sigma_i \langle \boldsymbol{v}_i, \boldsymbol{x} \rangle = \langle \boldsymbol{u}_i, \boldsymbol{b} \rangle$ and hence,

$$\langle \boldsymbol{v}_i, \boldsymbol{x} \rangle = \frac{\langle \boldsymbol{u}_i, \boldsymbol{b} \rangle}{\sigma_i}. \tag{6.10}$$

If we extend the nD vectors $\boldsymbol{v}_1$, ..., $\boldsymbol{v}_r$ to make an orthonormal basis $\{\boldsymbol{v}_1, ..., \boldsymbol{v}_r, \boldsymbol{v}_{r+1}, ..., \boldsymbol{v}_n\}$ of $\mathcal{R}^n$, we can expand $\boldsymbol{x}$ with respect to this basis in the form

$$\begin{aligned} \boldsymbol{x} &= \langle \boldsymbol{v}_1, \boldsymbol{x} \rangle \boldsymbol{v}_1 + \cdots + \langle \boldsymbol{v}_r, \boldsymbol{x} \rangle \boldsymbol{v}_r + \langle \boldsymbol{v}_{r+1}, \boldsymbol{x} \rangle \boldsymbol{v}_{r+1} + \cdots + \langle \boldsymbol{v}_n, \boldsymbol{x} \rangle \boldsymbol{v}_n \\ &= \frac{\langle \boldsymbol{u}_1, \boldsymbol{b} \rangle \boldsymbol{v}_1}{\sigma_1} + \cdots + \frac{\langle \boldsymbol{u}_r, \boldsymbol{b} \rangle \boldsymbol{v}_r}{\sigma_r} + \langle \boldsymbol{v}_{r+1}, \boldsymbol{x} \rangle \boldsymbol{v}_{r+1} + \cdots + \langle \boldsymbol{v}_n, \boldsymbol{x} \rangle \boldsymbol{v}_n. \end{aligned} \tag{6.11}$$

(↪ Appendix Eq. (A.35).) However, $\langle \boldsymbol{v}_{r+1}, \boldsymbol{x} \rangle$, ..., $\langle \boldsymbol{v}_n, \boldsymbol{x} \rangle$ are not determined.

Following the principle that we choose the solution that minimizes the square norm

$$\|\boldsymbol{x}\|^2 = \langle \boldsymbol{v}_1, \boldsymbol{x} \rangle^2 + \cdots + \langle \boldsymbol{v}_n, \boldsymbol{x} \rangle^2, \tag{6.12}$$

we adopt the solution for which $\langle \boldsymbol{v}_{r+1}, \boldsymbol{x} \rangle = \cdots = \langle \boldsymbol{v}_n, \boldsymbol{x} \rangle = 0$ (↪ Appendix Eq. (A.36)). As a result, $\boldsymbol{x}$ is expressed as follows:

$$\begin{aligned} \boldsymbol{x} &= \frac{\langle \boldsymbol{u}_1, \boldsymbol{b} \rangle \boldsymbol{v}_1}{\sigma_1} + \cdots + \frac{\langle \boldsymbol{u}_r, \boldsymbol{b} \rangle \boldsymbol{v}_r}{\sigma_r} = \frac{\boldsymbol{v}_1 \boldsymbol{u}_1^\top \boldsymbol{b}}{\sigma_1} + \cdots + \frac{\boldsymbol{v}_r \boldsymbol{u}_r^\top \boldsymbol{b}}{\sigma_r} \\ &= \Big(\frac{\boldsymbol{v}_1 \boldsymbol{u}_1^\top}{\sigma_1} + \cdots + \frac{\boldsymbol{v}_r \boldsymbol{u}_r^\top}{\sigma_r} \Big) \boldsymbol{b}. \end{aligned} \tag{6.13}$$

Namely, the least-squares solution $\boldsymbol{x}$ is given by

$$\boldsymbol{x} = \boldsymbol{A}^- \boldsymbol{b}. \tag{6.14}$$

Note that, as pointed out in the preceding chapter, we need to know the rank r of $\boldsymbol{A}$ for computing $\boldsymbol{A}^-$. It should be estimated from the fundamental principles or the fundamental laws behind the problem in question by inferring what the rank will be if the observation data are ideal. If no theoretical relationships or constraints among the n variables and m equations of Eq. (6.1) exist, we can let $r = \min(n, m)$. Otherwise, we compute the rank-constrained pseudoinverse, using the estimated rank r.

6.3 MULTIPLE EQUATIONS OF ONE VARIABLE

Consider, as an example, the simultaneous linear equations

$$a_1 x = b_1, \quad \ldots, \quad a_m x = b_m, \tag{6.15}$$

for $n = 1$, where we assume that $a_1, \ldots, a_m$ are not all 0. In vector form, it is written as

$$\boldsymbol{a}x = \boldsymbol{b}, \qquad \boldsymbol{a} = \begin{pmatrix} a_1 \\ \vdots \\ a_m \end{pmatrix} \ (\neq \boldsymbol{0}), \qquad \boldsymbol{b} = \begin{pmatrix} b_1 \\ \vdots \\ b_m \end{pmatrix}. \tag{6.16}$$

Since the pseudoinverse $\boldsymbol{a}^-$ of vector $\boldsymbol{a}$ is given by Eq. (5.10), the least-squares solution is given by

$$x = \boldsymbol{a}^-\boldsymbol{b} = \frac{\boldsymbol{a}^\top}{\|\boldsymbol{a}\|^2}\boldsymbol{b} = \frac{\langle \boldsymbol{a}, \boldsymbol{b} \rangle}{\|\boldsymbol{a}\|^2} = \frac{a_1 b_1 + \cdots + a_m b_m}{a_1^2 + \cdots + a_m^2}. \tag{6.17}$$

Formally, it is rewritten as

$$x = \frac{a_1^2(b_1/a_1) + \cdots + a_m^2(b_m/a_m)}{a_1^2 + \cdots + a_m^2}. \tag{6.18}$$

This is interpreted to be the weighted average of the individual solutions $x = b_i/a_i$ of Eq. (6.15) with weights a_i^2, where terms for $a_i = 0$ are ignored. It can be easily seen (↪ Problem 6.5) that this is the solution that minimizes the residual

$$J = (a_1 x - b_1)^2 + \cdots + (a_m x - b_m)^2. \tag{6.19}$$

6.4 SINGLE MULTIVARIATE EQUATION

For another example, consider a single linear equation

$$a_1 x_1 + \cdots + a_n x_n = b, \tag{6.20}$$

where we assume that $a_1, \ldots, a_n$ are not all 0. In vector form, it is written as

$$\langle \boldsymbol{a}, \boldsymbol{x} \rangle = b, \qquad \boldsymbol{a} = \begin{pmatrix} a_1 \\ \vdots \\ a_n \end{pmatrix} \ (\neq \boldsymbol{0}). \tag{6.21}$$

Since this is rewritten as $\boldsymbol{a}^\top \boldsymbol{x} = b$, the least-squares solution is given by $(\boldsymbol{a}^\top)^- \boldsymbol{x}$. As shown in Eq. (5.11), the pseudoinverse of a row vector $\boldsymbol{a}^\top$ is $\boldsymbol{a}^{-\top}$ $(= (\boldsymbol{a}^\top)^- = (\boldsymbol{a}^-)^\top)$. Hence, the least-squares solution is given by

$$\boldsymbol{x} = \boldsymbol{a}^{-\top} b = \frac{b\boldsymbol{a}}{\|\boldsymbol{a}\|^2}. \tag{6.22}$$

It is easy to see that $\langle \boldsymbol{a}, \boldsymbol{x} \rangle = b$ holds indeed. Since the ith component of Eq. (6.22) is $x_i = ba_i/\|\boldsymbol{a}\|^2$, we can write the ith term on the left side of Eq. (6.20) as

$$a_i x_i = \frac{a_i^2}{a_1^2 + \cdots + a_n^2} b. \tag{6.23}$$

This means that the n terms on the left side of Eq. (6.20) are portions of b on the right side distributed in proportion to the ratio $a_i^2 : \cdots : a_n^2$. It is easy to confirm that Eq. (6.22) is the solution that minimizes $\|\boldsymbol{x}\|^2$ subject to the condition that $\boldsymbol{x}$ satisfy Eq. (6.20) ($\hookrightarrow$ Problem 6.6).

6.5 GLOSSARY AND SUMMARY

least-squares method: Computing the solution $\boldsymbol{x}$ that minimizes the residual J of linear equations.

least-squares solution: The solution $\boldsymbol{x}$ that minimizes the residual J of linear equations. If multiple solutions exist, the one that minimizes the square norm $\|\boldsymbol{x}\|^2$ is chosen.

normal equation: The equation $\nabla_{\boldsymbol{x}} J = \boldsymbol{0}$ obtained by differentiating the residual J with respect to $\boldsymbol{x}$ and letting the result be $\boldsymbol{0}$.

residual: The sum of squares of the differences between the left and the right sides of linear equations. Formally, it is called the "residual sum of squares."

- The least-squares solution $\boldsymbol{x}$ of the linear equation $\boldsymbol{Ax} = \boldsymbol{b}$ satisfies the equation $\boldsymbol{Ax} = \boldsymbol{P}_{\mathcal{U}}\boldsymbol{b}$ obtained by replacing $\boldsymbol{b}$ by its projection $\boldsymbol{P}_{\mathcal{U}}\boldsymbol{b}$ onto the column domain $\mathcal{U}$ of $\boldsymbol{A}$ (Eq. (6.7)).
- The least-squares solution $\boldsymbol{x}$ of the linear equation $\boldsymbol{Ax} = \boldsymbol{b}$ is given by $\boldsymbol{x} = \boldsymbol{A}^-\boldsymbol{b}$ in terms of the pseudoinverse $\boldsymbol{A}^-$ of $\boldsymbol{A}$ (Eq. (6.13)).
- For a matrix whose elements are obtained from measurement data, we infer its rank by a theoretical consideration and compute the rank-constrained pseudoinverse.
- For computing the least-squares solution, we need not solve the normal equation.

6.6 SUPPLEMENTAL NOTES

In this chapter, we required the least-squares solution $\boldsymbol{x}$ to have minimum square norm $\|\boldsymbol{x}\|^2$ beside minimizing the residual J. What will happen if the minimum norm is not required? Evidently, the solution is not necessarily unique. This issue also has been studied in detail in the past.

If the minimum norm is not imposed, Eq. (6.11) implies that we can arbitrarily assign the values of $\langle \boldsymbol{v}_i, \boldsymbol{x} \rangle$, $i = r+1, ..., n$. This means that from Eqs. (6.13) and (6.14) we can write Eq. (6.11) as

$$\boldsymbol{x} = \boldsymbol{A}^{-}\boldsymbol{b} + c_{r+1}\boldsymbol{v}_{r+1} + \cdots + c_n\boldsymbol{v}_n, \tag{6.24}$$

for arbitrary c_r, ..., c_n, or equivalently

$$\boldsymbol{x} = \boldsymbol{A}^{-}\boldsymbol{b} + \boldsymbol{v}, \tag{6.25}$$

for arbitrary $\boldsymbol{v} \in \mathcal{V}^{\perp}$ (=ker($\boldsymbol{A}$)). Since $\boldsymbol{A}^{-}\boldsymbol{A}$ is the projection $\boldsymbol{P}_{\mathcal{V}}$ onto $\mathcal{V}$ ($\hookrightarrow$ Eq. (5.5)), the projection $\boldsymbol{P}_{\mathcal{V}^{\perp}}$ onto $\mathcal{V}^{\perp}$ is given by $\boldsymbol{I} - \boldsymbol{A}^{-}\boldsymbol{A}$. Hence, Eq. (6.25) is written as

$$\boldsymbol{x} = \boldsymbol{A}^{-}\boldsymbol{b} + (\boldsymbol{I} - \boldsymbol{A}^{-}\boldsymbol{A})\boldsymbol{v}, \tag{6.26}$$

for arbitrary $\boldsymbol{v} \in \mathcal{R}^n$.

This expression can be extended to cases in which $\boldsymbol{A}^{-}$ is not necessarily of the Moore–Penrose type. Let us write $\boldsymbol{A}^{+}$ for the Moore–Penrose type pseudoinverse and $\boldsymbol{A}^{-}$ for general (not necessarily Moore–Penrose type) pseudoinverses. From the reasoning that we obtained Eq. (6.26), we obtain

$$\boldsymbol{x} = \boldsymbol{A}^{+}\boldsymbol{b} + (\boldsymbol{I} - \boldsymbol{A}^{-}\boldsymbol{A})\boldsymbol{v}, \qquad \boldsymbol{v} \in \mathcal{R}^n \tag{6.27}$$

as long as $\boldsymbol{A}^{-}\boldsymbol{A}$ is the orthogonal projection of $\mathcal{R}^n$ onto $\mathcal{V}$.

As shown in Section 5.6, $\boldsymbol{A}^{-}\boldsymbol{A}$ defines for any pseudoinverse $\boldsymbol{A}^{-}$ a (not necessarily orthogonal) projection of $\mathcal{R}^n$ onto $\mathcal{V}$ ($\hookrightarrow$ Section 2.6). It is an orthogonal projection if and only if $\boldsymbol{A}^{-}\boldsymbol{A}$ is symmetric ($(\boldsymbol{A}^{-}\boldsymbol{A})^{\top} = \boldsymbol{A}^{-}\boldsymbol{A}$) ($\hookrightarrow$ Problems 2.4 and 2.7). Such a pseudoinverse is called *minimum norm pseudoinverse*. Thus, we conclude that Eq. (6.27) holds for all minimum norm pseudoinverses.

6.7 PROBLEMS

6.1. Show that if $m > n$ and if the columns of $\boldsymbol{A}$ are linearly independent, i.e., $r = n$, then

(1) the least-squares solution $\boldsymbol{x}$ is given by

$$\boldsymbol{x} = (\boldsymbol{A}^{\top}\boldsymbol{A})^{-1}\boldsymbol{A}^{\top}\boldsymbol{b}, \tag{6.28}$$

(2) and the residual J is written in the form

$$J = \|\boldsymbol{b}\|^2 - \langle \boldsymbol{A}\boldsymbol{x}, \boldsymbol{b} \rangle. \tag{6.29}$$

6.2. Show that if $m > n = r$, the following identity holds:

$$(\boldsymbol{A}^{\top}\boldsymbol{A})^{-1}\boldsymbol{A}^{\top} = \boldsymbol{A}^{-}. \tag{6.30}$$

6.3. Show that if $n > m$ and if the rows of $\boldsymbol{A}$ are linearly independent, i.e., $r = m$, then the residual J is 0 and the least-squares solution $\boldsymbol{x}$ is given by

$$\boldsymbol{x} = \boldsymbol{A}^\top(\boldsymbol{A}\boldsymbol{A}^\top)^{-1}\boldsymbol{b}. \tag{6.31}$$

6.4. Show that if $n > m = r$, the following identity holds:

$$\boldsymbol{A}^\top(\boldsymbol{A}\boldsymbol{A}^\top)^{-1} = \boldsymbol{A}^-. \tag{6.32}$$

6.5. Show that the solution x given by Eq. (6.17) minimizes the sum of square of Eq. (6.19).

6.6. Show that Eq. (6.22) minimizes $\|\boldsymbol{x}\|^2$ over all $\boldsymbol{x}$ that satisfy Eq. (6.20).

CHAPTER 7

Probability Distribution of Vectors

In this chapter, we regard measurement data that conain noise not as definitive values but as "random variables" specified by probability distributions. The principal parameters that characterize a probability distribution are the "mean" (average) and the "covariance matrix." In particular, the "normal" (or "Gaussian") distribution is characterized by the mean and the covariance matrix alone. We show that if the probability is not distributed over the entire space but is restricted to some domain, e.g., constrained to be on a planar surface or on a sphere, the covariance matrix becomes singular. In such a case, the probability distribution is characterized by the pseudoinverse of the covariance matrix. We illustrate how this leads to a practical method for comparing computational accuracy of such data.

7.1 COVARIANCE MATRICES OF ERRORS

A vector $\boldsymbol{x}$ is a *random variable* if its value is not deterministic but is specified by some (or assumed) *probability distribution*. All observations in real situations are definitive values; regarding them, or "modeling" them as random variables by introducing an artificial probability distribution is only a mathematical convenience. However, this can very well approximate the reality in many engineering problems where sensor measurement involves some uncertainties. Hence, viewing data as random variables provides a very useful analytical tool in practical applications.

To be specific, we interpret an observed value $\boldsymbol{x}$ to be the sum of its true value $\bar{\boldsymbol{x}}$ (a definitive value) and a noise term $\Delta\boldsymbol{x}$ (a random variable):

$$\boldsymbol{x} = \bar{\boldsymbol{x}} + \Delta\boldsymbol{x}. \tag{7.1}$$

We usually assume that the noise term $\Delta\boldsymbol{x}$ has *expectation* (or average) $\mathbf{0}$; if it has an expectation not equal to $\mathbf{0}$, we can model its probability distribution after subtracting it. Thus, we assume

$$E[\Delta\boldsymbol{x}] = \mathbf{0}, \tag{7.2}$$

where $E[\cdot]$ denotes expectation over the probability distribution of the noise. We define the *covariance matrix* (↪ Problem 7.1) of the noise by

$$\boldsymbol{\Sigma} = E[\Delta\boldsymbol{x}\Delta\boldsymbol{x}^\top]. \tag{7.3}$$

From this, we see that $\boldsymbol{\Sigma}$ is a positive semidefinite matrix, i.e., its eigenvalues are positive or zero (↪ Problem 7.2). The *mean square* of the noise term $\Delta \boldsymbol{x}$ is given by

$$E[\|\Delta \boldsymbol{x}\|^2] = \mathrm{tr}\boldsymbol{\Sigma}, \tag{7.4}$$

i.e., the trace of the covariance matrix $\boldsymbol{\Sigma}$ (↪ Problem 7.3).

Let σ_1^2, ..., σ_n^2 be the nonnegative eigenvalues of the covariance matrix $\boldsymbol{\Sigma}$, and $\boldsymbol{u}_1$, ..., $\boldsymbol{u}_n$ the orthonormal system of the corresponding unit eigenvectors. Then, $\boldsymbol{\Sigma}$ has the following spectral decomposition (↪ Eq. (3.3)):

$$\boldsymbol{\Sigma} = \sum_{i=1}^{n} \sigma_i^2 \boldsymbol{u}_i \boldsymbol{u}_i^\top. \tag{7.5}$$

We call the directions of the vectors $\boldsymbol{u}_1$, ..., $\boldsymbol{u}_n$ the *principal axes* of the noise distribution. The values σ_1^2, ..., σ_n^2 indicate the variance of the noise in the respective directions; σ_1, ..., σ_n are their standard deviations. In fact, the magnitude of the noise term $\Delta \boldsymbol{x}$ along $\boldsymbol{u}_i$, i.e., the projected length in the direction of $\boldsymbol{u}_i$ is $\langle \Delta \boldsymbol{x}, \boldsymbol{u}_i \rangle$ (↪ Eq. (2.18)), and its mean square is

$$E[\langle \Delta \boldsymbol{x}, \boldsymbol{u}_i \rangle^2] = E[(\boldsymbol{u}_i^\top \Delta \boldsymbol{x})(\Delta \boldsymbol{x}^\top \boldsymbol{u}_i)] = \langle \boldsymbol{u}_i, E[\Delta \boldsymbol{x} \Delta \boldsymbol{x}^\top] \boldsymbol{u}_i \rangle = \langle \boldsymbol{u}_i, \boldsymbol{\Sigma} \boldsymbol{u}_i \rangle = \sigma_i^2. \tag{7.6}$$

If all the eigenvalues are equal, i.e., $\sigma_1^2 = \cdots = \sigma_n^2$ $(= \sigma^2)$, the noise is said to be *isotropic*, in which case the noise occurrence is equally likely in all directions and the covariance matrix $\boldsymbol{\Sigma}$ of Eq. (7.3) reduces to

$$\boldsymbol{\Sigma} = \sigma^2 \sum_{i=1}^{n} \boldsymbol{u}_i \boldsymbol{u}_i^\top = \sigma^2 \boldsymbol{I}. \tag{7.7}$$

(↪ Eq. (3.4).) Otherwise, the noise is *anisotropic*; the likelihood of the noise occurence depends on the direction. In particular, the eigenvector $\boldsymbol{u}_{\max}$ for the maximum eigenvalue $\sigma_{\max}^2$ is the direction along which the noise is most likely to occur; $\sigma_{\max}^2$ is the variance in that direction. This is easily seen if we note that the projected length of the noise term $\Delta \boldsymbol{x}$ onto the direction of a unit vector $\boldsymbol{u}$ is given by $\langle \Delta \boldsymbol{x}, \boldsymbol{u} \rangle$ (↪ Section 2.4) and that its square mean is, as shown in Eq. (3.4), $E[\langle \Delta \boldsymbol{x}, \boldsymbol{u} \rangle^2] = \langle \boldsymbol{u}, \boldsymbol{\Sigma} \boldsymbol{u} \rangle$, which is a quadratic form of the symmetric matrix $\boldsymbol{\Sigma}$. Hence, the unit vector $\boldsymbol{u}$ that maximizes it is given by the unit eigenvector $\boldsymbol{u}_{\max}$ for the maximum eigenvalue $\sigma_{\max}$ (↪ Appendix A.10).

If there exists an eigenvector $\boldsymbol{u}_i$ for which the eigenvalue is 0, we infer that no noise occurs in that direction. In practice, this means that the variation of $\boldsymbol{x}$ in that direction is physically prohibited.

7.2 NORMAL DISTRIBUTION OF VECTORS

A typical probability distribution is the *normal* (or *Gaussian*) *distribution* introduced by Gauss (see Section 6.1 about him). In the nD space $\mathcal{R}^n$, it is specified by the expectation $\bar{\boldsymbol{x}}$ and the

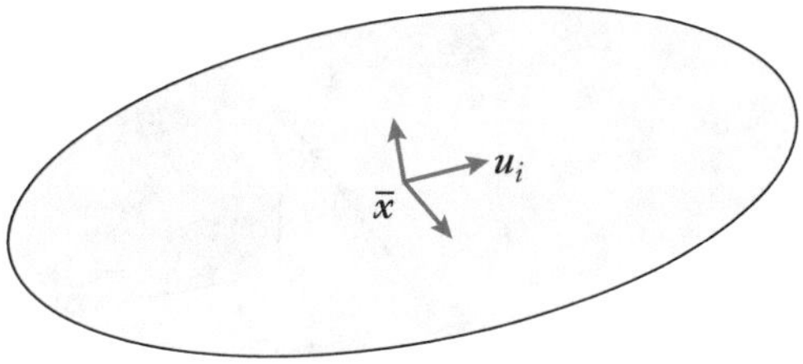

Figure 7.1: Error ellipsoid. It is centered on the expectation $\bar{\boldsymbol{x}}$, and the principal axes $\boldsymbol{u}_i$ of the covariance matrix $\boldsymbol{\Sigma}$ are its axes of symmetry. The radius along each principal axis is the standard deviation σ_i of errors in that direction.

covariance matrix $\boldsymbol{\Sigma}$ alone. Its distribution density has the form

$$p(\boldsymbol{x}) = C \exp(-\frac{1}{2}\langle \boldsymbol{x} - \bar{\boldsymbol{x}}, \boldsymbol{\Sigma}^{-1}(\boldsymbol{x} - \bar{\boldsymbol{x}})\rangle), \tag{7.8}$$

where C is a normalization constant determined so that the integration over the entire space $\mathcal{R}^n$ equals 1 (to be precise, $C = 1/\sqrt{(2\pi)^n \sigma_1^2 \cdots \sigma_n^2}$). The distribution extends to infinity, and the covariance matrix $\boldsymbol{\Sigma}$ is assumed to be positive definite (i.e., all the eigenvalues are positive). It satisfies

$$\int_{\mathcal{R}^n} p(\boldsymbol{x}) d\boldsymbol{x} = 1, \qquad \int_{\mathcal{R}^n} \boldsymbol{x}\, p(\boldsymbol{x}) d\boldsymbol{x} = \bar{\boldsymbol{x}},$$
$$\int_{\mathcal{R}^n} (\boldsymbol{x} - \bar{\boldsymbol{x}})(\boldsymbol{x} - \bar{\boldsymbol{x}})^\top p(\boldsymbol{x}) d\boldsymbol{x} = \boldsymbol{\Sigma}, \tag{7.9}$$

where $\int_{\mathcal{R}^n} (\cdots) d\boldsymbol{x}$ denotes integration over the entire $\mathcal{R}^n$.

The surface on which the probability density $p(\boldsymbol{x})$ equals 1

$$\langle \boldsymbol{x} - \bar{\boldsymbol{x}}, \boldsymbol{\Sigma}^{-1}(\boldsymbol{x} - \bar{\boldsymbol{x}})\rangle = 1 \tag{7.10}$$

is called the *error ellipsoid*. (Fig. 7.1). This is an ellipsoid centered on $\bar{\boldsymbol{x}}$; it is also called the *error ellipse* for two variables and the *confidence interval* for a single variable. Each of the eigenvectors of the covariance matrix $\boldsymbol{\Sigma}$ is the axis of symmetry, and the radius in that direction is the standard deviation σ_i in that direction ($\hookrightarrow$ Problems 7.4 and 7.5). Thus, the error ellipse visualizes directional dependence of the likelihood of error occurrence.

In some engineering applications, including computer graphics and computer vision, we often consider covariance matrices which are not positive definite. The fact that the covariance matrix has eigenvalue 0 means that the disturbance in the corresponding eigenvector is prohibited. Suppose, for example, we want to evaluate the uncertainty of the position of a particular point on the display surface or in the image. Then, the noise disturbance occurs only in 2D, and the perpendicular displacement is prohibited. It seems then that it is sufficient to define a 2D coordinate system in that plane and consider a normal distribution of two variables. However,

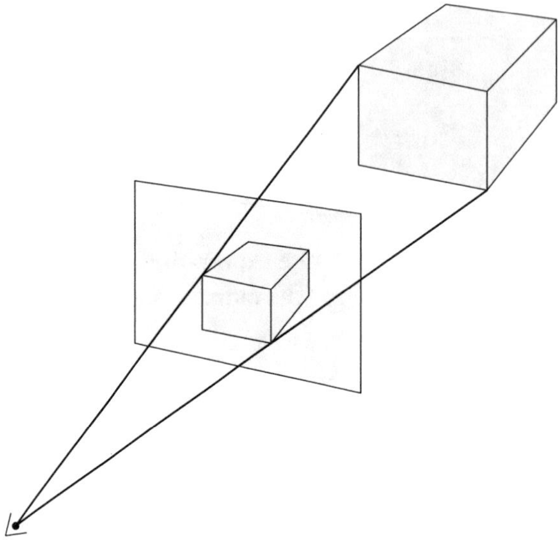

Figure 7.2: The figures and patterns seen on the display or in the image are interpreted to be perspective images of some 3D objects viewed from a particular viewpoint.

it is often more convenient to regard that plane as a surface in the 3D space. This is because we usually define a particular point, called the *viewpoint*, in the 3D scene, which corresponds to the position of the human eye or the camera lens center. We identify the display surface or the image as the camera image plane and analyze the geometric properties of figures or patterns on it by regarding them as perspective views of objects in the 3D scene (Fig. 7.2).

Considering such applications, let r be the rank of the $n \times n$ covariance matrix $\boldsymbol{\Sigma}$, and $\boldsymbol{u}_1, ..., \boldsymbol{u}_r$ the orthonormal system of its unit eigenvectors for positive eigenvalues $\sigma_1^2, ..., \sigma_r^2$ (> 0). Let $\mathcal{U}$ be the rD subspace they span. In other words, we are assuming that all deviations occur within $\mathcal{U}$ and that no displacements are allowed in the direction of $\mathcal{U}^\perp$. In this case, the probability density of the normal distribution with expectation $\bar{\boldsymbol{x}}$ and covariance matrix $\boldsymbol{\Sigma}$ has the form

$$p(\boldsymbol{x}) = C \exp(-\frac{1}{2}\langle \boldsymbol{x} - \bar{\boldsymbol{x}}, \boldsymbol{\Sigma}^-(\boldsymbol{x} - \bar{\boldsymbol{x}})\rangle), \tag{7.11}$$

where $\boldsymbol{\Sigma}^-$ is the pseudoinverse of the covariance matrix $\boldsymbol{\Sigma}$, and C is a normalization constant determined so that the integration over $\mathcal{U}$, not over the entire $\mathcal{R}^n$, equals 1 (to be precise, $C = 1/\sqrt{(2\pi)^r \sigma_1^2 \cdots \sigma_r^2}$). Then, we obtain the relationships, corresponding to Eqs. (7.9),

$$\int_{\mathcal{U}} p(\boldsymbol{x}) d\boldsymbol{x} = 1, \qquad \int_{\mathcal{U}} \boldsymbol{x} p(\boldsymbol{x}) d\boldsymbol{x} = \bar{\boldsymbol{x}},$$

$$\int_{\mathcal{U}} (\boldsymbol{x} - \bar{\boldsymbol{x}})(\boldsymbol{x} - \bar{\boldsymbol{x}})^\top p(\boldsymbol{x}) d\boldsymbol{x} = \boldsymbol{\Sigma}, \tag{7.12}$$

where $\int_{\mathcal{U}}(\cdots)d\boldsymbol{x}$ denotes integration over $\mathcal{U}$.

The covariance matrix $\boldsymbol{\Sigma}$ and its pseudoinverse $\boldsymbol{\Sigma}^-$ have the following spectral decompositions:

$$\boldsymbol{\Sigma} = \sum_{i=1}^{r} \sigma_i^2 \boldsymbol{u}_i \boldsymbol{u}_i^\top, \qquad \boldsymbol{\Sigma}^- = \sum_{i=1}^{r} \frac{\boldsymbol{u}_i \boldsymbol{u}_i^\top}{\sigma_i^2}. \tag{7.13}$$

Hence, the projection matrix $\boldsymbol{P}_{\mathcal{U}}$ $(= \sum_{i=1}^{r} \boldsymbol{u}_i \boldsymbol{u}_i^\top)$ onto the subspace $\mathcal{U}$ satisfies

$$\boldsymbol{P}_{\mathcal{U}}\boldsymbol{\Sigma} = \boldsymbol{\Sigma}\boldsymbol{P}_{\mathcal{U}} = \boldsymbol{P}_{\mathcal{U}}\boldsymbol{\Sigma}\boldsymbol{P}_{\mathcal{U}} = \boldsymbol{\Sigma},$$

$$\boldsymbol{P}_{\mathcal{U}}\boldsymbol{\Sigma}^- = \boldsymbol{\Sigma}^-\boldsymbol{P}_{\mathcal{U}} = \boldsymbol{P}_{\mathcal{U}}\boldsymbol{\Sigma}^-\boldsymbol{P}_{\mathcal{U}} = \boldsymbol{\Sigma}^-. \tag{7.14}$$

It follows that we obtain ($\hookrightarrow$ Appendix Eq. (A.27))

$$\begin{aligned} \langle \boldsymbol{x} - \bar{\boldsymbol{x}}, \boldsymbol{\Sigma}^-(\boldsymbol{x} - \bar{\boldsymbol{x}})\rangle &= \langle \boldsymbol{x} - \bar{\boldsymbol{x}}, (\boldsymbol{P}_{\mathcal{U}}\boldsymbol{\Sigma}^-\boldsymbol{P}_{\mathcal{U}})(\boldsymbol{x} - \bar{\boldsymbol{x}})\rangle \\ &= \langle \boldsymbol{P}_{\mathcal{U}}(\boldsymbol{x} - \bar{\boldsymbol{x}}), \boldsymbol{\Sigma}^-(\boldsymbol{P}_{\mathcal{U}}(\boldsymbol{x} - \bar{\boldsymbol{x}}))\rangle. \end{aligned} \tag{7.15}$$

In other words, the use of the pseudoinverse $\boldsymbol{\Sigma}^-$ is equivalent to considering the normal distribution not of $\boldsymbol{x} - \bar{\boldsymbol{x}}$ but of its projection $\boldsymbol{P}_{\mathcal{U}}(\boldsymbol{x} - \bar{\boldsymbol{x}})$ onto $\mathcal{U}$. Within the subspace $\mathcal{U}$, the distribution is regarded as a normal distribution with variance σ_i^2 (> 0) in each principal axis direction, defining a covariance matrix that is positive definite within $\mathcal{U}$.

EXAMPLE: NORMAL DISTRIBUTION OVER THE IMAGE PLANE

The normal distribution over the image plane is represented as follows. Let $(\bar{x}, \bar{y})$ be the true position of an observed point (x, y), and write $(x, y) = (\bar{x} + \Delta x, \bar{y} + \Delta y)$. We assume that the noise terms Δx and Δy are subject to a normal distribution of expectation 0 and variance/covariance specified by

$$E[(\Delta x)^2] = \sigma_x^2, \qquad E[(\Delta y)^2] = \sigma_y^2, \qquad E[(\Delta x)(\Delta y)] = \gamma. \tag{7.16}$$

We regard the image plane, or the display surface, as the plane $z = 1$ in the 3D space (Fig. 7.3) and represent points (x, y) and $(\bar{x}, \bar{y})$ on it by 3D vectors

$$\boldsymbol{x} = \begin{pmatrix} x \\ y \\ 1 \end{pmatrix}, \qquad \bar{\boldsymbol{x}} = \begin{pmatrix} \bar{x} \\ \bar{y} \\ 1 \end{pmatrix}. \tag{7.17}$$

Representing points in 2D by 3D vectors this way is equivalent to the use of *homogeneous coordinates* well known in projective geometry. Then, the probability density of $\boldsymbol{x}$ has the form

$$p(\boldsymbol{x}) = C \exp(-\frac{1}{2}\langle \boldsymbol{x} - \bar{\boldsymbol{x}}, \boldsymbol{\Sigma}^-(\boldsymbol{x} - \bar{\boldsymbol{x}})\rangle), \qquad C = \frac{1}{2\pi(\sigma_x^2\sigma_y^2 - \gamma^2)}, \tag{7.18}$$

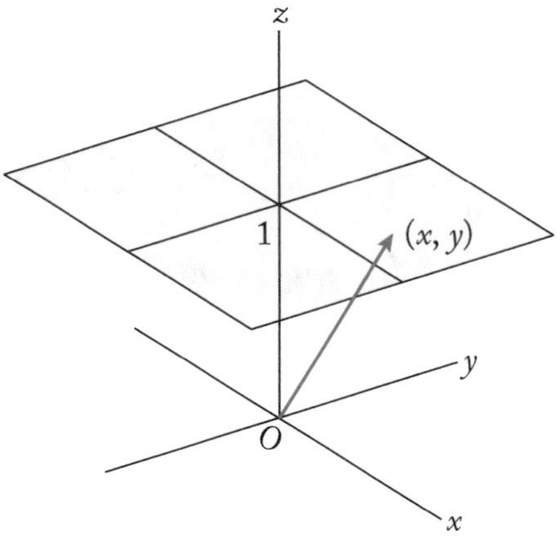

Figure 7.3: The image plane or the display is interpreted to be the plane $z = 1$ in 3D, and the position (x, y) on it is represented by a 3D vector $\boldsymbol{x}$.

where the covariance matrix $\boldsymbol{\Sigma}$ is singular and has the form

$$\boldsymbol{\Sigma} = \begin{pmatrix} \sigma_x^2 & \gamma & 0 \\ \gamma & \sigma_y^2 & 0 \\ 0 & 0 & 0 \end{pmatrix}. \tag{7.19}$$

7.3 PROBABILITY DISTRIBUTION OVER A SPHERE

Applications which involve non-positive definite covariance matrices, other than distributions over planes, include distributions over a "sphere." In some problems of physics and engineering, sensor data are directional, i.e., only orientations can be measured. If we normalize the direction vector to unit norm, each observed datum can be regarded as a point over a unit sphere.

A typical example of directional data is observations using cameras. A camera can identify the directions of incoming rays of light, but the depth, i.e., the distance to the object, is unknown. This limitation cannot be resolved even if we use multiple cameras, or equivalently move one camera. In fact, if we move a camera over a short distance relative to a small object nearby or over a long distance relative to a large object in the distance, the observed image is the same. Today, various computer vision techniques are established for reconstructing the shape of 3D scenes and objects using camera images, but the absolute scale of the reconstructed shape is always indeterminate. This scale indeterminacy is not limited to the reconstructed shape. Many different types of matrix that characterize the structure of the scene can be computed from images, but such matrices are usually determined up to scale; typical examples include the "fundamental matrix" and the "homography matrix" (see Section 7.5 for these). In practice, they are normalized

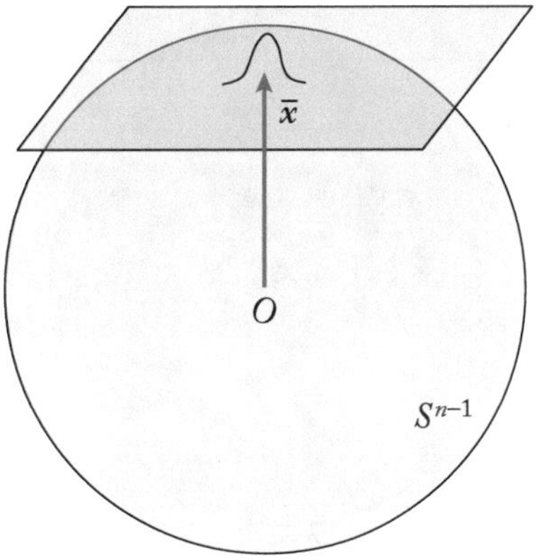

Figure 7.4: The measurement $\boldsymbol{x}$ on a sphere is thought of as distributed over a small region of the tangent plane around its expectation $\bar{\boldsymbol{x}}$.

so that the sum of the squares of the elements is 1. If we view an $n \times n$ matrix as an n^2D vector consisting of the n^2 matrix elements, it can be regarded as a point on a unit sphere in the n^2D space after normalization.

Considering these, we assume that the measured value $\boldsymbol{x}$ is an nD unit vector. It is identified with a point on the $(n-1)$D unit sphere S^{n-1} in the nD space $\mathcal{R}^n$, and a probability distribution is defined around its true position $\bar{\boldsymbol{x}}$. However, mathematical analysis is generally very difficult. First of all, the normal distribution, which is the most common distribution, cannot be defined, because the sphere S^{n-1} is of finite size while the normal distribution extends infinitely over the entire space $\mathcal{R}^n$. In practical problems, however, today's sensors including cameras are highly accurate so that the noise component $\Delta\boldsymbol{x}$ of $\boldsymbol{x}$ is usually very small compared with $\boldsymbol{x}$. For example, the accuracy in locating a particular point in an image using an image processing algorithm is usually around 1~3 pixels. Consequently, the distribution of $\boldsymbol{x}$ is thought to be limited to a small region surrounding $\bar{\boldsymbol{x}}$ on the sphere S^{n-1}. Hence, we can view this as a distribution over the tangent plane to S^{n-1} at $\bar{\boldsymbol{x}}$ (Fig. 7.4). We can then define the expectation and the covariance matrix of $\boldsymbol{x}$, thereby a normal distribution of $\boldsymbol{x}$. The tangent plane to S^{n-1} at $\bar{\boldsymbol{x}}$ is an $(n-1)$D (hyper-)plane, whose unit surface normal is $\bar{\boldsymbol{x}}$ itself. Hence, from Eq. (2.17) the projection matrix onto the tangent plane is given by

$$\boldsymbol{P}_{\bar{\boldsymbol{x}}} = \boldsymbol{I} - \bar{\boldsymbol{x}}\bar{\boldsymbol{x}}^\top. \tag{7.20}$$

Consider, as a practical application, the reliability evaluation of the measurement $\boldsymbol{x}$. Suppose we repeat the measurement and observe $\boldsymbol{x}_1, \ldots, \boldsymbol{x}_N$. Alternatively, suppose you want to evaluate the performance of a computational procedure for computing $\boldsymbol{x}$. You artificially add random noise to the original observations and compute $\boldsymbol{x}$ from the noisy data. Let $\boldsymbol{x}_1, \ldots, \boldsymbol{x}_N$ the results for different noise. In either case, you want to evaluate how $\boldsymbol{x}_1, \ldots, \boldsymbol{x}_N$ differ from the theoretical value $\bar{\boldsymbol{x}}$. The standard procedure for statistical treatment is computing the mean

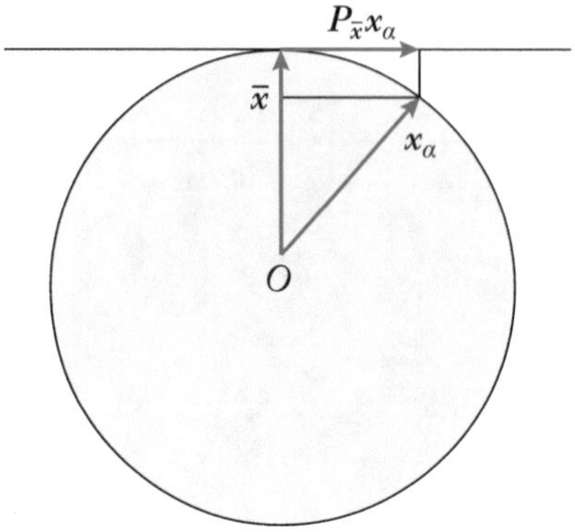

Figure 7.5: The deviation of a measurement $\boldsymbol{x}_\alpha$ from its expectation $\bar{\boldsymbol{x}}$ is evaluated by the projection $\boldsymbol{P}_{\bar{\boldsymbol{x}}}\boldsymbol{x}_\alpha$ of $\boldsymbol{x}_\alpha$ onto the tangent plane at $\bar{\boldsymbol{x}}$.

and the covariance matrix. However, arithmetic average of points on a sphere is no longer on that sphere, and covariance computation is not obvious, either. To circumvent this difficulty, we project the data onto the tangent plane, assuming that noise is very small.

Let $\boldsymbol{x}_\alpha \in S^{n-1}$ be the αth measurement. Its projection onto the tangent plane (Fig. 7.5) is given by

$$\hat{\boldsymbol{x}}_\alpha = \boldsymbol{P}_{\bar{\boldsymbol{x}}}\boldsymbol{x}_\alpha. \tag{7.21}$$

The *sample mean* $\boldsymbol{m}$ and the *sample covariance matrix* $\boldsymbol{S}$ are computed by

$$\boldsymbol{m} = \frac{1}{N}\sum_{\alpha=1}^{N}\hat{\boldsymbol{x}}_\alpha, \qquad \boldsymbol{S} = \frac{1}{N}\sum_{\alpha=1}^{N}(\hat{\boldsymbol{x}}_\alpha - \boldsymbol{m})(\hat{\boldsymbol{x}}_\alpha - \boldsymbol{m})^\top, \tag{7.22}$$

where the expression "sample $\cdots$" means replacing the integration in the expectation computations with respect to the true probability distribution, as in Eqs. (7.10) and (7.12), by the arithmetic average over all "realizations," i.e., actual measurements or observations.

The sample mean $\boldsymbol{m}$ indicates the average deviation from the true position $\bar{\boldsymbol{x}}$ on the tangent plane. Its magnitude $\|\boldsymbol{m}\|$, which is ideally 0, is called the *bias*. The diagonal element S_{ii} of the sample covariance matrix $\boldsymbol{S}$ is the sample variance of the ith component $\hat{x}_{i\alpha}$ of $\hat{\boldsymbol{x}}_\alpha$, and the non-diagonal element S_{ij}, $i \neq j$ is the sample covariance of $\hat{x}_{i\alpha}$ and $\hat{x}_{j\alpha}$ ($\hookrightarrow$ Problem 7.6). In statistics, the term "bias" usually means the deviation of the expectation of a random variable from the true value. If it is 0, the random variable is said to be *unbiased*. For realizations, the bias means the deviation of the sample mean from its true value. Here, we are using this term in the realization sense.

The square root of the trace of the sample covariance matrix $\boldsymbol{S}$

$$\sqrt{\mathrm{tr}\boldsymbol{S}} = \sqrt{\frac{1}{N}\sum_{\alpha=1}^{N} \|\hat{\boldsymbol{x}}_\alpha - \boldsymbol{m}\|^2} \tag{7.23}$$

is called the *root-mean-square error*, or the *RMS error* for short, which is a typical indicator of the observation accuracy or computational performance

From the definition of $\boldsymbol{P}_{\bar{\boldsymbol{x}}}$ of Eq. (7.21), the sample covariance matrix $\boldsymbol{S}$ satisfies

$$\boldsymbol{P}_{\bar{\boldsymbol{x}}}\boldsymbol{S} = \boldsymbol{S}\boldsymbol{P}_{\bar{\boldsymbol{x}}} = \boldsymbol{P}_{\bar{\boldsymbol{x}}}\boldsymbol{S}\boldsymbol{P}_{\bar{\boldsymbol{x}}} = \boldsymbol{S}, \tag{7.24}$$

and has rank $n - 1$. If the distribution of $\boldsymbol{x}$ is regarded as a normal distribution, its *empirical probability density* is given by

$$p(\boldsymbol{x}) = C\exp(-\frac{1}{2}\langle \boldsymbol{x} - \boldsymbol{m}, \boldsymbol{S}^-(\boldsymbol{x} - \boldsymbol{m})\rangle), \tag{7.25}$$

where C is the normalization constant determined so that integration over the tangent plane is 1. The expression "empirical $\cdots$" means replacing the parameters included in the theoretical expression by values estimated from realizations like the sample mean and the sample covariance matrix.

7.4 GLOSSARY AND SUMMARY

bias: The deviation of the expectation of a random variable from its true value, or the deviation of the sample mean from the true value.

covariance matrix: The matrix given for a random variable $\boldsymbol{x}$ so that its ith diagonal element is the variance of the ith component and the (i, j) element is the covariance of the ith and the jth components.

empirical probability density: For a probability density involving unknown parameters that specify the probability distribution, the parameter values can be estimated from observed data occurrences, e.g., by computing sample means. If the estimated parameter values are substituted, the resulting density function is called the "empirical probability density."

error ellipsoid: The ellipsoid that describes the region in which the occurrence of the values of a random variable is highly likely. It is called the "error ellipse" for two variables and the "confidence interval" for a single variable.

expectation: The average of the possible values of a random variable weighted by the probabilities of their occurrences. For a continuous random variable, it is given by the integral after multiplied by the probability density.

isotropy: A probability distribution of a random variable is "isotropic" if the characteristics of occurrence are the same for all directions. If they depend on the direction, the distribution is "anisotropic."

mean square: The expectation of the square norm of a random variable $\boldsymbol{x}$.

normal distribution: The probability distribution modeled by the German mathematician Gauss (↪ Section 6.1). Its density function has the form of exponential of the quadratic form of the random variable. Physicists often call it the "Gaussian distribution."

principal axis: The direction of each eigenvector of the covariance matrix.

probability distribution: A function that specifies the range of possible values of a variable and their likelihoods.

random variable: A variable whose value is not definitive but specified by a probability distribution.

root-mean-square error: The square root of the average of the square differences between the occurrences and the sample mean. It is a typical index for measuring the error magnitude and often abbreviated as the "RMS error."

sample mean: The approximation of the expectation $E[\boldsymbol{x}]$ of random variable $\boldsymbol{x}$ by the algebraic average $(1/N)\sum_{\alpha=1}^{N}\boldsymbol{x}_\alpha$ of its occurrences $\boldsymbol{x}_\alpha$, $\alpha = 1, ..., N$. For a covariance matrix, we obtain the corresponding "sample covariance matrix" by replacing the expectation operation by sample averaging.

viewpoint: For computer graphics and computer vision, the figures and patterns seen on the display or in the image are interpreted to be perspective images of some 3D objects viewed from a particular point in the 3D space. That point is called the "viewpoint" and corresponds to the position of the human eye or the lens center of the camera.

- If the covariance matrix $\boldsymbol{\Sigma}$ of a random variable $\boldsymbol{x}$ has the spectral decomposition $\boldsymbol{\Sigma} = \sum_{i=1}^{n}\sigma_i^2\boldsymbol{u}_i\boldsymbol{u}_i^\top$, the variance of the error in the direction of the ith principal axis $\boldsymbol{u}_i$ is σ_i^2.

- The eigenvector $\boldsymbol{u}_{\max}$ of the covariance matrix $\boldsymbol{\Sigma}$ for the largest eigenvalue $\sigma_{\max}^2$ is in the direction along which the error is most likely to occur, and $\sigma_{\max}^2$ is the variance in that direction.

- If the distribution is isotropic, the covariance matrix $\boldsymbol{\Sigma}$ is a scalar multiple of the identity matrix $\boldsymbol{I}$.

- The normal distribution of $\boldsymbol{x}$ is specified by its expectation $\bar{\boldsymbol{x}}$ and the inverse $\boldsymbol{\Sigma}^{-1}$ of its covariance matrix $\boldsymbol{\Sigma}$ (Eq. (7.8)).
- If the distribution of $\boldsymbol{x}$ is constrained in $\mathcal{R}^n$ so that the error cannot occur in certain directions, its covariance matrix $\boldsymbol{\Sigma}$ is a singular matrix whose rank is less than n. If the distribution is normal, it is specified by its expectation $\bar{\boldsymbol{x}}$ and the pseudoinverse $\boldsymbol{\Sigma}^-$ (Eq. (7.11)).
- Specifying a normal distribution using the pseudoinverse $\boldsymbol{\Sigma}^-$ means considering the normal distribution of the projection $\boldsymbol{P}_{\mathcal{U}}(\boldsymbol{x} - \bar{\boldsymbol{x}})$ of $\boldsymbol{x} - \bar{\boldsymbol{x}}$ onto the column domain $\mathcal{U}$ of $\boldsymbol{\Sigma}$; due to this constraint, no error occurs in the directions $\mathcal{U}^{\perp}$ orthogonal to $\mathcal{U}$.
- We regard those measurement data for which only their directions are obtained and their absolute scales are indeterminate as random variables that distribute over the unit sphere S^{n-1} in $\mathcal{R}^n$. They include most of the values computed from images in computer vision applications.
- The distribution of $\boldsymbol{x}$ over S^{n-1} is regarded as occurring in the tangent plane to S^{n-1} at $\bar{\boldsymbol{x}}$, if $\boldsymbol{x}$ is concentrated on a very small region around its true value $\bar{\boldsymbol{x}}$. This is the case for most quantities measured using ordinary sensors or computed from image data,
- When errors are small, we can evaluate the properties of the values $\boldsymbol{x}_\alpha$, $\alpha = 1, ..., N$, observed on S^{n-1} by projecting them onto the tangent plane at $\bar{\boldsymbol{x}}$ using the projection matrix $\boldsymbol{P}_{\bar{\boldsymbol{x}}}$.

7.5 SUPPLEMENTAL NOTES

Regarding sensor measurement as a random process has been a standard analytical tool in signal processing for communication and control, and various mathematical theory have been introduced, including Wiener filters and Kalman filters, based on covariance matrices, correlation functions, and power spectrum (some other topics will be discussed in the next chapter).

Such a statistical approach has also been introduced to computer vision research for geometric analysis of images and 3D reconstruction from image data. Analysis of computer vision is based on *perspective projection* of 3D scenes depicted in Fig. (7.2). One of the earliest work in the computer vision context was Kanatani [8]. This is perhaps the first consistent work of introducing singular covariance matrices for analyzing geometrically constrained data on planes and spheres. Since then various optimization techniques for geometric computation have been extensively investigated. See [5, 12] for geometric computation for 3D reconstruction from images.

As mentioned in Section 7.3, many computer vision problems reduce to analysis of matrices that are determined only up to scale. A typical example is the 3×3 *fundamental matrix* that relates the 3D scene structure to two camera images; analyzing 3D using two (or more) cameras is called *stereo vision*. If we compute the fundamental matrix from the relative configuration of the two cameras and their internal parameters, we can reconstruct the 3D structure from two stereo images in terms of the fundamental matrix (see [12] for reconstruction procedure). It has scale indeterminacy, and hence the reconstructed 3D shape also has scale indeterminacy.

On the other hand, if we view a planar surface in the scene by two cameras, the corresponding images are related by a transformation called *homography*, or 2D *projective transformation*, specified by a 3×3 matrix, which we call the *homography matrix*. If we compute the homography matrix from two images of the same planar surface in the scene, we can reconstruct its 3D structure (see [12] for reconstruction procedure). However, the reconstruction is only up to scale, since the homography matrix can be determined only up to scale.

Yet another example is *ellipse fitting*. Circular objects in the scene are projected onto the camera image as ellipses, so detecting ellipses in images is one of the basic processes of computer vision. An ellipse is represented as a 3×3 matrix consisting of the coefficients of the ellipse equation. From the detected ellipse equation, we can compute the 3D position and orientation of the circular object that we are viewing (see [11] for reconstruction procedure). However, the 3×3 ellipse matrix has scale indeterminacy, and the absolute scale of the computed circular object is also indeterminate.

Various numerical techniques have been proposed for computing the fundamental matrix, the homography matrix, and the ellipse matrix. All such matrices are regarded as points on a unit sphere in 9D, and the accuracy and reliability of computation is evaluated by projecting them onto the tangent plane to the unit sphere, using the projection matrix. Their uncertainty distribution is modeled as a normal distribution involving a singular covariance matrix and its pseudoinverse (see [11, 12] for more details).

7.6 PROBLEMS

7.1. Show that if we let $\boldsymbol{x} = \begin{pmatrix} x_i \end{pmatrix}$, the diagonal element Σ_{ii} of the covariance matrix $\boldsymbol{\Sigma}$ of Eq. (7.3) gives the variance of x_i and that the non-diagonal element Σ_{ij}, $i \neq j$ gives the covariance of x_i and x_j.

7.2. Show that the matrix $\boldsymbol{X} = \boldsymbol{x}\boldsymbol{x}^\top$ defined from a vector $\boldsymbol{x}$ is a positive semidefinite symmetric matrix, i.e., a symmetric matrix whose eigenvalues are positive or 0. Also show that this is the case for the matrix $\boldsymbol{X} = \sum_{\alpha=1}^{N} \boldsymbol{x}_\alpha \boldsymbol{x}_\alpha^\top$ defined by multiple vectors $\boldsymbol{x}_1$, ..., $\boldsymbol{x}_N$, too.

7.3. Show that $\mathrm{tr}(\boldsymbol{x}\boldsymbol{x}^\top) = \|\boldsymbol{x}\|^2$ holds for any vector $\boldsymbol{x}$. Also show that $\mathrm{tr}(\sum_{\alpha=1}^{N} \boldsymbol{x}_\alpha \boldsymbol{x}_\alpha^\top) = \sum_{\alpha=1}^{N} \|\boldsymbol{x}_\alpha\|^2$ holds for any multiple vectors $\boldsymbol{x}_1$, ..., $\boldsymbol{x}_N$, too.

7.4. Write down explicitly the surface of Eq. (7.10) in 3D when $\boldsymbol{\Sigma}$ is a diagonal matrix.

7.5. Show that the ellipsoid given by Eq. (7.10) has its center at the expectation $\bar{\boldsymbol{x}}$ with the eigenvectors $\boldsymbol{u}_i$ of the covariance matrix $\boldsymbol{\Sigma}$ as its axes of symmetry and that the radius in each directions is the standard deviation σ_i of the error in that direction.

7.6. Write $\hat{\boldsymbol{x}}_\alpha = \left(\hat{x}_{i\alpha}\right)$, and show that the diagonal element S_{ii} of the sample covariance matrix $\boldsymbol{S}$ of Eq. (7.22) is the variance of $\hat{x}_{i\alpha}$ and that its non-diagonal element S_{ij}, $i \neq j$ is the sample covariance of $x_{i\alpha}$ and $x_{j\alpha}$.

CHAPTER 8

Fitting Spaces

In this chapter, we generalize line fitting in 2D and plane fitting in 3D to general dimensions and consider how to fit subspaces and affine spaces to a given set of points in nD. A subspace is a space spanned by vectors starting from the origin, and an "affine space" is a translation of a subspace to a general position. The fitting is done hierarchically: we first fit a lower dimensional space, starting from a 0D space (= a point), then determine a space with an additional dimension so that the discrepancy is minimized, and continue this. This principle corresponds to what is known as "Karhunen–Loéve expansion" in signal processing and pattern recognition and as "principal component analysis" in statistics. The fitted space is computed from the spectral decomposition of a matrix, which we call the "covariance matrix," but also can be obtained from its singular value decomposition. We point out that the use of the singular value decomposition is more efficient with smaller computational complexity.

8.1 FITTING SUBSPACES

Given N points $\boldsymbol{x}_1$, ..., $\boldsymbol{x}_N$ in nD, we consider the problem of finding an rD subspace that is the closest to them, where we assume that $N \geq r$. If $n = 3$ and $r = 1$, for example, this is the problem of line fitting: we want to find a line passing through the origin that is as close to the given N points as possible. For $n = 3$ and $r = 2$, this is plane fitting: we compute a plane passing through the origin that is close to the given N points (Fig. 8.1). Here, the "closeness" is measured by the sum of square distances.

Finding a subspace is equivalent to finding its basis. Let $\boldsymbol{u}_1$, ..., $\boldsymbol{u}_r$ be an orthonormal basis of the rD subspace $\mathcal{U}$ to be fitted, and let $\{\boldsymbol{u}_1, ..., \boldsymbol{u}_n\}$ be its extension to an orthonormal basis of the entire $\mathcal{R}^n$. Put another way, the problem is to find an orthonormal basis $\{\boldsymbol{u}_i\}$ of $\mathcal{R}^n$ such that the subspace $\mathcal{U}$ spanned by its first r vectors $\boldsymbol{u}_1$, ..., $\boldsymbol{u}_r$ is as close to the given N points as possible.

The distance of each point $\boldsymbol{x}_\alpha$ to the subspace $\mathcal{U}$ equals the length of the rejection $\boldsymbol{P}_{\mathcal{U}^\perp}\boldsymbol{x}_\alpha$ from $\mathcal{U}$ (Fig. 8.2), where $\boldsymbol{P}_{\mathcal{U}^\perp} = \sum_{i=r+1}^{n} \boldsymbol{u}_i\boldsymbol{u}_i^\top$ is the projection matrix onto the orthogonal complement $\mathcal{U}^\perp$ of $\mathcal{U}$ ($\hookrightarrow$ Eq. (2.9)). Hence, the sum of square distances of the N points $\boldsymbol{x}_1$, ..., $\boldsymbol{x}_N$ from the subspace $\mathcal{U}$ is

$$J = \sum_{\alpha=1}^{N} \|\boldsymbol{P}_{\mathcal{U}^\perp}\boldsymbol{x}_\alpha\|^2. \tag{8.1}$$

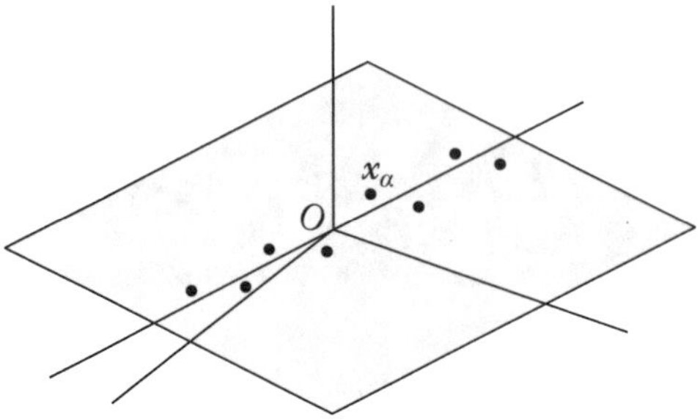

Figure 8.1: Fitting a line passing through the origin O (1D subspace) or a plane passing through the origin O (2D subspace) to a given set of points $\{\boldsymbol{x}_\alpha\}$, $\alpha = 1, \ldots, N$, so that all the points are close to the fitted space.

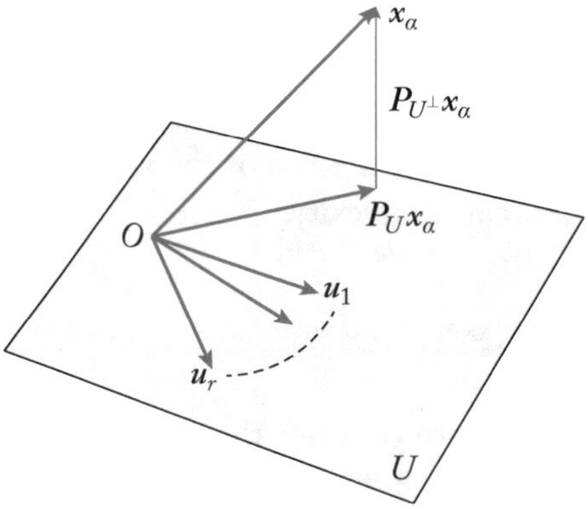

Figure 8.2: The distance of each point $\boldsymbol{x}_\alpha$ to subspace $\mathcal{U}$ equals the length of the rejection $\boldsymbol{P}_{\mathcal{U}^\perp}\boldsymbol{x}_\alpha$ from $\mathcal{U}$.

From Eq. (2.12), this is rewritten as

$$J = \sum_{\alpha=1}^{N}(\|\boldsymbol{x}_\alpha\|^2 - \|\boldsymbol{P}_{\mathcal{U}}\boldsymbol{x}_\alpha\|^2) = \sum_{\alpha=1}^{N}\|\boldsymbol{x}_\alpha\|^2 - \sum_{\alpha=1}^{N}\|\boldsymbol{P}_{\mathcal{U}}\boldsymbol{x}_\alpha\|^2. \tag{8.2}$$

Hence, minimizing the sum of squares of Eq. (8.1) is equivalent to maximizing the sum of square projected lengths onto the subspace $\mathcal{U}$

$$K = \sum_{\alpha=1}^{N}\|\boldsymbol{P}_{\mathcal{U}}\boldsymbol{x}_\alpha\|^2. \tag{8.3}$$

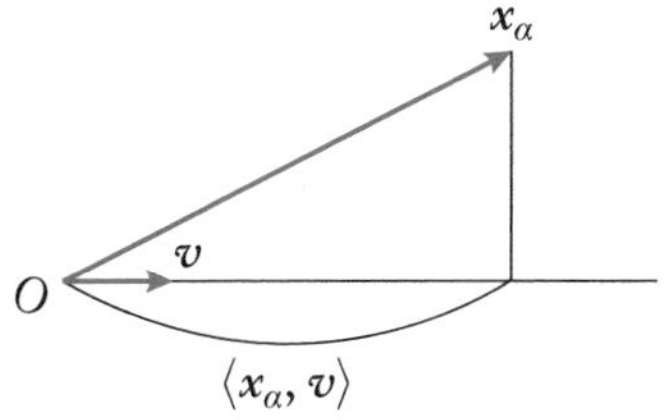

Figure 8.3: The projected length of point $\boldsymbol{x}_\alpha$ onto a 1D subspace, i.e., a line, with a unit vector $\boldsymbol{v}$ as its basis equals $\langle \boldsymbol{x}_\alpha, \boldsymbol{v}\rangle$.

8.2 HIERARCHICAL FITTING

First, consider the problem of fitting a 1D subspace. Let $\boldsymbol{v}$ (a unit vector) be its basis (Fig. 8.3). The projected length of each point $\boldsymbol{x}_\alpha$ onto that direction is $\langle \boldsymbol{x}_\alpha, \boldsymbol{v}\rangle$ ($\hookrightarrow$ Eq. (2.16)). Hence, the sum of their squares over all points is

$$K = \sum_{\alpha=1}^{N} \langle \boldsymbol{x}_\alpha, \boldsymbol{v}\rangle^2 = \sum_{\alpha=1}^{N} \boldsymbol{v}^\top \boldsymbol{x}_\alpha \boldsymbol{x}_\alpha^\top \boldsymbol{v} = \langle \boldsymbol{v}, \Big(\sum_{\alpha=1}^{N} \boldsymbol{x}_\alpha \boldsymbol{x}_\alpha^\top\Big)\boldsymbol{v}\rangle = \langle \boldsymbol{v}, \boldsymbol{\Sigma}\boldsymbol{v}\rangle, \tag{8.4}$$

where we define the $n \times n$ matrix $\boldsymbol{\Sigma}$ by

$$\boldsymbol{\Sigma} = \sum_{\alpha=1}^{N} \boldsymbol{x}_\alpha \boldsymbol{x}_\alpha^\top. \tag{8.5}$$

This matrix is called by many different names, including the "moment matrix" and the "scatter matrix," both borrowed from physics. In the following, we call it the *covariance matrix*, borrowing from statistics, at the risk of possible confusion but for convenience' sake. In statistics, this is, if divided by N, equal to the sample covariance matrix of the N sample data $\boldsymbol{x}_\alpha$ around the origin, by regarding the origin as the mean ($\hookrightarrow$ Eq. (7.22)).

Equation (8.4) is a quadratic form of a symmetric matrix $\boldsymbol{\Sigma}$ in $\boldsymbol{v}$. The vector $\boldsymbol{v}$ that maximizes this is the unit eigenvector of the matrix $\boldsymbol{\Sigma}$ for the maximum eigenvalue, and the resulting value of K equals that maximum eigenvalue of $\boldsymbol{\Sigma}$ ($\hookrightarrow$ Appendix A.10). By construction, $\boldsymbol{\Sigma}$ is a positive semidefinite symmetric matrix, and its eigenvalues are all nonnegative ($\hookrightarrow$ Problem 7.2). If we let $\sigma_1^2 \geq \cdots \geq \sigma_n^2 \geq 0$ be its eigenvalues, $\boldsymbol{\Sigma}$ has the following spectral decomposition ($\hookrightarrow$ Eq. (3.3)):

$$\boldsymbol{\Sigma} = \sigma_1^2 \boldsymbol{u}_1 \boldsymbol{u}_1^\top + \cdots + \sigma_n^2 \boldsymbol{u}_n \boldsymbol{u}_n^\top, \qquad \sigma_1^2 \geq \cdots \geq \sigma_n^2 \geq 0. \tag{8.6}$$

From this observation, we conclude that the basis of the 1D subspace $\mathcal{U}_1$ that best fits to the point set $\{\boldsymbol{x}_\alpha\}$, $\alpha = 1, ..., N$, is given by $\boldsymbol{v} = \boldsymbol{u}_1$ and that the resulting sum of the square projected lengths equals σ_1^2.

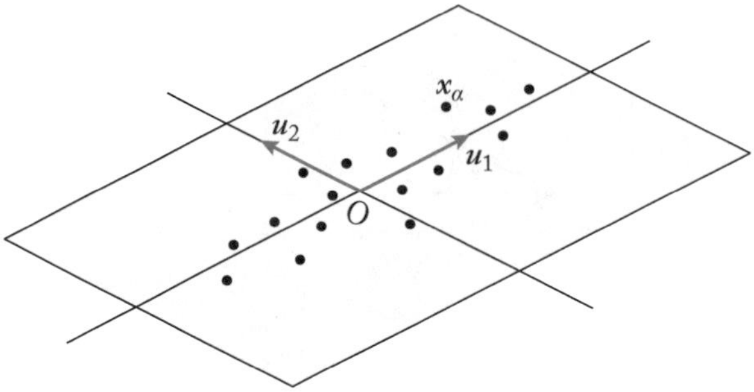

Figure 8.4: The vector $\boldsymbol{u}_1$ indicates the direction in which the spread of the N points $\{\boldsymbol{x}_\alpha\}$, $\alpha = 1, \ldots, N$, is the largest. The vector $\boldsymbol{u}_2$, for which among the directions orthogonal to $\boldsymbol{u}_1$ the square sum of the projected length onto it is the largest, indicates the direction in which the spread of $\{\boldsymbol{x}_\alpha\}$ is the second largest.

The obtained direction $\boldsymbol{u}_1$ indicates the orientation in which the N points $\{\boldsymbol{x}_\alpha\}$ spread to the largest extent (Fig. 8.4). If the set $\{\boldsymbol{x}_\alpha\}$ approximately spreads linearly, its distribution is well approximated by a line in the direction of $\boldsymbol{u}_1$. However, if the points spread in other directions, too, this approximation is not sufficient. So, we want to find a direction $\boldsymbol{v}$ that is orthogonal to $\boldsymbol{u}_1$ along which the sum of the square projected lengths is maximized. The square projected length onto the direction of $\boldsymbol{v}$ is again given by Eq. (8.4). The unit vector $\boldsymbol{v}$ that maximizes it subject to the condition $\langle \boldsymbol{v}, \boldsymbol{u}_1 \rangle = 0$ is given by $\boldsymbol{u}_2$ of the spectral decomposition Eq. (8.6) of $\boldsymbol{\Sigma}$, and the corresponding value of K equals σ_2^2 ($\hookrightarrow$ Appendix A.10). The vector $\boldsymbol{u}_2$ indicates the direction in which the spread of the N points $\{\boldsymbol{x}_\alpha\}$ is the second largest.

From this observation, we conclude that the basis of the 2D subspace $\mathcal{U}_2$ that best fits to $\{\boldsymbol{x}_\alpha\}$ is given by $\boldsymbol{u}_1$ and $\boldsymbol{u}_2$. Since $\boldsymbol{u}_1$ and $\boldsymbol{u}_2$ are mutually orthogonal, the sum of the square projected lengths K onto $\mathcal{U}$ is the sum of the corresponding eigenvalues for $\boldsymbol{u}_1$ and $\boldsymbol{u}_2$, namely $K = \sigma_1^2 + \sigma_2^2$.

By the same argument, we see that the direction in which the spread is the third largest is given by $\boldsymbol{u}_3$ in Eq. (8.6); $\boldsymbol{u}_1$, $\boldsymbol{u}_2$, and $\boldsymbol{u}_3$ span the 3D subspace $\mathcal{U}_3$ that best fits to $\{\boldsymbol{x}_\alpha\}$, and the sum of the projected lengths is $K = \sigma_1^2 + \sigma_2^2 + \sigma_3^2$. Repeating this argument, we see that the basis of the rD subspace $\mathcal{U}_r$ that best fits to $\{\boldsymbol{x}_\alpha\}$ is $\boldsymbol{u}_1, \ldots, \boldsymbol{u}_r$ and that the sum of the square projected lengths is $K = \sigma_1^2 + \cdots + \sigma_r^2$.

From Eqs. (8.2) and (8.3), we obtain $J = \sum_{\alpha=1}^{N} \|\boldsymbol{x}_\alpha\|^2 - K$. From Eq. (8.5), we see that $\mathrm{tr}\boldsymbol{\Sigma} = \sum_{\alpha=1}^{N} \|\boldsymbol{x}_\alpha\|^2$ ($\hookrightarrow$ Problem 7.3), and from Eq. (8.6) we obtain ($\hookrightarrow$ Problem 8.1)

$$\mathrm{tr}\boldsymbol{\Sigma} = \sigma_1^2 + \cdots + \sigma_n^2. \tag{8.7}$$

Hence, the sum of square distance to $\mathcal{U}_r$ is

$$J = \sigma_{r+1}^2 + \cdots + \sigma_n^2, \tag{8.8}$$

which we call the *residual sum of squares* or the *residual* for short. Since $\sigma_1^2 \geq \cdots \geq \sigma_n^2$, we obtain a smaller residual J as we increase the dimension r of the fitting.

8.3 FITTING BY SINGULAR VALUE DECOMPOSITION

The argument of the preceding section shows that an rD subspace $\mathcal{U}_r$ that best fits to N points $\{\boldsymbol{x}_\alpha\}$, $\alpha = 1, ..., N$, is obtained by first computing the covariance matrix $\boldsymbol{\Sigma}$ of Eq. (8.5) and then computing its spectral decomposition in the form of Eq. (8.6).

On the other hand, consider the $n \times N$ matrix

$$\boldsymbol{X} = \begin{pmatrix} \boldsymbol{x}_1 & \cdots & \boldsymbol{x}_N \end{pmatrix}, \tag{8.9}$$

consisting of vectors $\boldsymbol{x}_\alpha$, $\alpha = 1, ..., N$, as its columns. Then, the covariance matrix $\boldsymbol{\Sigma}$ of Eq. (8.5) is given by

$$\boldsymbol{\Sigma} = \boldsymbol{X}\boldsymbol{X}^\top. \tag{8.10}$$

Since we are assuming that $N \geq n$, the singular value decomposition of $\boldsymbol{X}$ is

$$\boldsymbol{X} = \sigma_1 \boldsymbol{u}_1 \boldsymbol{v}_1^\top + \cdots + \sigma_n \boldsymbol{u}_n \boldsymbol{v}_n^\top, \qquad \sigma_1 \geq \cdots \geq \sigma_n \geq 0, \tag{8.11}$$

because the eigenvalues of $\boldsymbol{\Sigma} = \boldsymbol{X}\boldsymbol{X}^\top$ equals the square eigenvalues of $\boldsymbol{X}$ ($\hookrightarrow$ Eq. (4.2)). Moreover, the singular vectors $\boldsymbol{u}_i$ and $\boldsymbol{v}_i$ are, respectively, the eigenvectors of $\boldsymbol{X}\boldsymbol{X}^\top$ and $\boldsymbol{X}^\top\boldsymbol{X}$. It follows that for fitting an rD subspace $\mathcal{U}_r$ to N points $\{\boldsymbol{x}_\alpha\}$, $\alpha = 1, ..., N$, we may alternatively compute the singular value decomposition of the matrix $\boldsymbol{X}$ of Eq. (8.9) defined by the N points. Then, the left singular vectors $\boldsymbol{u}_1, ..., \boldsymbol{u}_r$ provide the basis of $\mathcal{U}_r$, and the residual is given by $J = \sum_{i=r+1}^{n} \sigma_i^2$.

Thus, the use of the spectral decomposition and the use of the singular value decomposition both give the same result. In actual applications, however, we should use the singular value decomposition. This is for the sake of computational efficiency. Computations involving matrices and vectors consist of computations of "sums of products." Computing the sum of n products requires n multiplications and $n - 1$ additions/subtractions. Disregarding the term -1, we can view the number of multiplications and the number of additions/subtractions as approximately equal. For complexity analysis, therefore, it is sufficient to evaluate the number of multiplications. For spectral decomposition, we first compute the covariance matrix $\boldsymbol{\Sigma}$ by Eq. (8.10), which requires n^2N multiplication (this is the same if Eq. (8.5) is used). The complexity of the spectral decomposition, i.e., the computation of eigenvalues and eigenvectors, of an $n \times n$ may differ from algorithm to algorithm but is approximately n^3. Hence, the total complexity of computing $\boldsymbol{\Sigma}$ and its spectral decomposition is approximately $n^2(N + n)$. On the other hand, the complexity of the singular value decomposition of an $n \times N$ matrix is approximately n^2N for $n \leq N$ and approximately nN^2 for $N \leq n$. Hence, the singular value decomposition is overwhelmingly efficient when $N \ll n$. Even if for $n \leq N$, we can save nearly equivalent time for eigenvalue

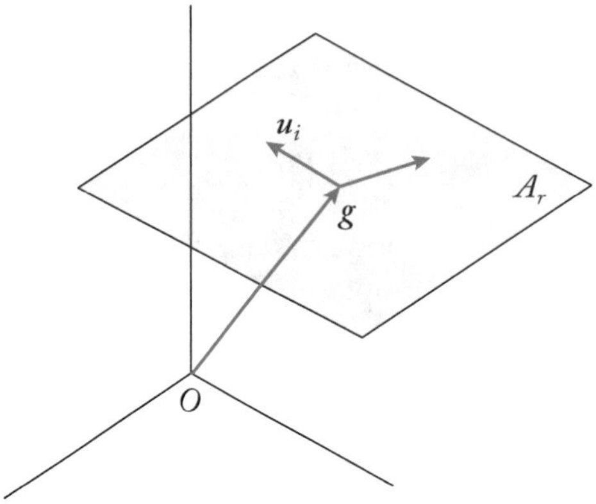

Figure 8.5: An rD affine space $\mathcal{A}_r$ is the space spanned by r linearly independent vectors $\{\boldsymbol{u}_i\}$, $i = 1, ..., r$, starting from a point $\boldsymbol{g}$ in $\mathcal{R}^n$.

and eigenvector computation of the covariance matrix (see Section 8.6 for implications of this in video analysis).

This time saving is often underestimated. Pattern information processing involves a large amount of data, and usually iterations are required for accuracy improvement. It is not uncommon, for instance, that the number of point data extracted from multiple images is as large as hundreds of thousands. In such a case, several hours of computation can sometimes be reduced to several seconds by simply replacing spectral decomposition computation by singular value computation.

8.4 FITTING AFFINE SPACES

Subspace fitting is a generalization of fitting to a point sequence a line passing through the origin and a plane passing through the origin. In practice, however, we often need to fit a line that does not pass through the origin and a plane that does not pass through the origin. This type of fitting is generalized to "affine space fitting." An *affine space* is a translation of a subspace. An rD affine space $\mathcal{A}_r$ is specified by a point $\boldsymbol{g}$ in nD and r linearly independent vectors $\boldsymbol{u}_1, ..., \boldsymbol{u}_r$ starting from it; it is the set of all points in the form

$$\boldsymbol{x} = \boldsymbol{g} + c_1\boldsymbol{u}_1 + \cdots + c_r\boldsymbol{u}_r, \tag{8.12}$$

for arbitrary $c_1, ..., c_r$ (Fig. 8.5). Without losing generality, we can let the basis $\{\boldsymbol{u}_i\}$, $i = 1, ..., r$, be an orthonormal system. Instead of specifying a point and r directions, we may alternatively define an rD affine space by specifying $r + 1$ points in nD in such a way that whichever point of it is regarded as the origin O, the remaining r points span an rD subspace. Such $r + 1$ points are said be *in general position* ($\hookrightarrow$ Problem 8.2).

Given N points $\{\boldsymbol{x}_\alpha\}$, $\alpha = 1, ..., N$, in $\mathcal{R}^n$, we consider the problem of finding an rD affine space $\mathcal{A}_r$ that approximates them, where we assume $N \geq r + 1$. The affine space $\mathcal{A}_r$ is specified by a point $\boldsymbol{g}$ it passes through and an orthonormal basis $\{\boldsymbol{u}_i\}$ starting from it.

First, we need to specify the point $\boldsymbol{g}$. This task can be regarded as fitting an 0D affine space (= one point) to the N points $\{\boldsymbol{x}_\alpha\}$. So, we choose $\boldsymbol{g}$ so that it minimizes the sum of square distances $\sum_{\alpha=1}^{N} \|\boldsymbol{x}_\alpha - \boldsymbol{g}\|^2$. Such a point is given by the centroid

$$\boldsymbol{g} = \frac{1}{N}\sum_{\alpha=1}^{N} \boldsymbol{x}_\alpha \tag{8.13}$$

of the N points $\{\boldsymbol{x}_\alpha\}$ ($\hookrightarrow$ Problem 8.3).

Once $\boldsymbol{g}$ is determined, all we need to do is, according to the argument in the preceding sections, fit an rD subspace to vectors $\{\boldsymbol{x}_\alpha - \boldsymbol{g}\}$, $\alpha = 1, ..., N$, regarding g as the origin. Namely, we compute the covariance matrix

$$\boldsymbol{\Sigma} = \sum_{\alpha=1}^{N} (\boldsymbol{x}_\alpha - \boldsymbol{g})(\boldsymbol{x}_\alpha - \boldsymbol{g})^\top, \tag{8.14}$$

around $\boldsymbol{g}$ ($\hookrightarrow$ Problem 8.4). If its spectral decomposition is written in the form of Eq. (8.6), the eigenvectors $\boldsymbol{u}_1, ..., \boldsymbol{u}_r$ span the affine space $\mathcal{A}_r$ around $\boldsymbol{g}$. The residual, i.e., the sum of square distances of individual points $\boldsymbol{x}_\alpha$ to $\mathcal{A}_r$, is given by $K = \sigma_{r+1}^2 + \cdots + \sigma_n^2$.

However, as pointed out in the preceding section, it is more efficient to compute the singular value decomposition (8.11) of the matrix

$$\boldsymbol{X} = \begin{pmatrix} \boldsymbol{x}_1 - \boldsymbol{g} & \cdots & \boldsymbol{x}_N - \boldsymbol{g} \end{pmatrix} \tag{8.15}$$

consisting of data points, *without computing the covariance matrix*, and then obtain the basis vectors $\boldsymbol{u}_1, ..., \boldsymbol{u}_r$.

The technique of computing the spectral decomposition and the singular value decomposition of the covariance matrix of Eq. (8.14) and hierarchically fitting rD affine spaces $\mathcal{A}_r$ is known as the *Karhunen–Loéve expansion*, or the *KL-expansion* for short, in signal and pattern recognition applications and *principal component analysis* in statistics. Using these, we can compress data and images for efficient transmission by omitting insignificant high dimensional terms. We can also do statistical prediction and testing from multidimensional statistical data. See Section 8.6 for more discussions about these.

EXAMPLE: LINE FITTING IN 2D

We want to fit a line to N points $(x_1, y_1), ..., (x_N, y_N)$ given in the plane, First, we compute the centroid $(g_x, g_y) = \sum_{\alpha=1}^{N}(x_\alpha, y_\alpha)/N$ and define the matrix

$$\boldsymbol{X} = \begin{pmatrix} x_1 - g_x & \cdots & x_N - g_x \\ y_1 - g_y & \cdots & y_N - g_y \end{pmatrix}. \tag{8.16}$$

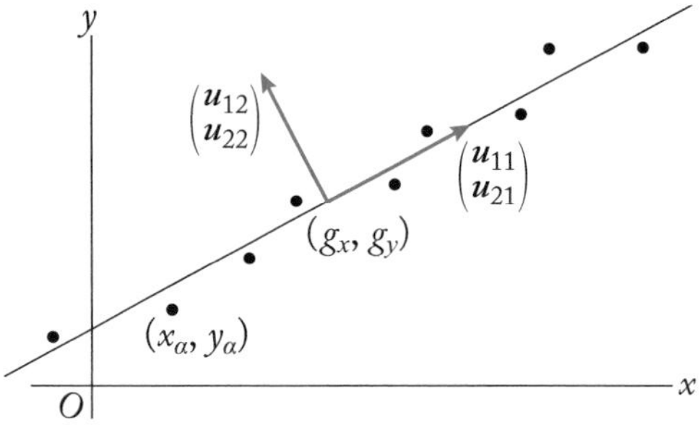

Figure 8.6: Fitting a line to N points $\{(x_\alpha, y_\alpha)\}$, $\alpha = 1, ..., N$.

Its singular value decomposition has the form

$$X = \sigma_1 \begin{pmatrix} u_{11} \\ u_{21} \end{pmatrix} \begin{pmatrix} v_{11} & \cdots & v_{N1} \end{pmatrix} + \sigma_2 \begin{pmatrix} u_{12} \\ u_{22} \end{pmatrix} \begin{pmatrix} v_{12} & \cdots & v_{N2} \end{pmatrix}. \tag{8.17}$$

This indicates that the line to be fitted passes through (g_x, g_y) and extends in the direction of $(u_{11}, u_{21})^\top$ (Fig. 8.6). The equation of the fitted line is given by

$$u_{12}(x - g_x) + u_{22}(y - g_y) = 0. \tag{8.18}$$

EXAMPLE: PLANE FITTING IN 3D

We want to fit a plane to N points (x_1, y_1, z_1), ..., (x_N, y_N, z_N) given in the space. First, we compute the centroid $(g_x, g_y, g_z) = \sum_{\alpha=1}^{N}(x_\alpha, y_\alpha, z_\alpha)/N$ and define the matrix

$$X = \begin{pmatrix} x_1 - g_x & \cdots & x_N - g_x \\ y_1 - g_y & \cdots & y_N - g_y \\ z_1 - g_z & \cdots & z_N - g_z \end{pmatrix}. \tag{8.19}$$

Its singular decomposition has the form

$$\begin{aligned} X = {} & \sigma_1 \begin{pmatrix} u_{11} \\ u_{21} \\ u_{31} \end{pmatrix} \begin{pmatrix} v_{11} & \cdots & v_{N1} \end{pmatrix} + \sigma_2 \begin{pmatrix} u_{12} \\ u_{22} \\ u_{32} \end{pmatrix} \begin{pmatrix} v_{12} & \cdots & v_{N2} \end{pmatrix} \\ & + \sigma_3 \begin{pmatrix} u_{13} \\ u_{23} \\ u_{33} \end{pmatrix} \begin{pmatrix} v_{13} & \cdots & v_{N3} \end{pmatrix}. \end{aligned} \tag{8.20}$$

This indicates that the plane to be fitted passes through (g_x, g_y, g_z) and extends in the directions of $(u_{11}, u_{21}, u_{31})^\top$ and $(u_{12}, u_{22}, u_{32})^\top$ (Fig. 8.7). The vector $(u_{13}, u_{23}, u_{33})^\top$, which is orthogonal to both of them, is the unit surface normal, and the equation of the plane is given by

$$u_{13}(x - g_x) + u_{23}(y - g_y) + u_{33}(x - g_z) = 0. \tag{8.21}$$

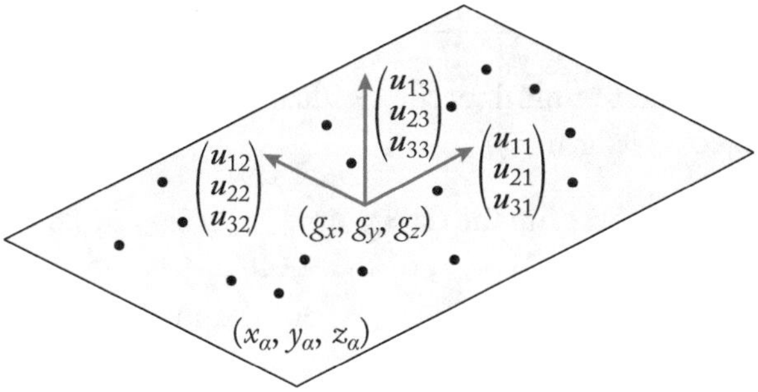

Figure 8.7: Plane fitting. Fitting a plane to N points $\{(x_\alpha, y_\alpha, z_\alpha)\}$, $\alpha = 1, \ldots, N$.

8.5 GLOSSARY AND SUMMARY

affine space: Translation of a subspace. Namely, a linear space spanned by r ($\leq n$) linearly independent vectors starting from a point in $\mathcal{R}^n$. It is a generalization of lines and planes.

covariance matrix: The matrix defined by $\boldsymbol{\Sigma} = \sum_{\alpha=1}^{N} \boldsymbol{x}_\alpha \boldsymbol{x}_\alpha^\top$ for data $\boldsymbol{x}_\alpha$, $\alpha = 1, \ldots, N$; it is a terminology borrowed from statistics. Also called the "moment matrix" or the "scatter matrix," terminologies, borrowed from physics.

general position: A configuration of $r + 1$ ($\leq n + 1$) points in $\mathcal{R}^n$ such that if one is regarded as the origin O, the remaining r span an rD subspace.

Karhunen–Loéve expansion: Constructing an orthogonal basis from a set of signals or patterns for signal processing and pattern recognition and then expanding individual signals or patterns with respect to the constructed basis. It is abbreviated to "KL expansion." This is applied to data and image compression for improving the efficiency of transmission and display by omitting those basis vectors which have small contributions. Mathematically, the principle is reduced to fitting affine spaces.

principal component analysis: Constructing mutually orthogonal axes, or principal axes, that characterize a large number of multi-dimensional statistical data in statistics. The obtained axes are used for understanding the general tendency of the statistical data, extracting a small number of statistics that characterize the data well, and doing prediction and testing. Mathematically, the principle is reduced to fitting affine spaces.

residual: The sum of square distances of the data points $\boldsymbol{x}_\alpha$, $\alpha = 1, \ldots, N$, from the fitted space (subspace or affine space). Also called the "residual sum of squares." It serves as a measure of goodness of fit (the smaller, the better).

- Fitting a subspace that minimizes the residual is equivalent to maximizing the sum of the square projected lengths.

- An rD subspace that best fits the data points $\{\boldsymbol{x}_\alpha\}$, $\alpha = 1, ..., N$, is the space spanned by the eigenvectors $\boldsymbol{u}_i$, $i = 1, ..., r$, of the spectral decomposition $\boldsymbol{\Sigma} = \sum_{i=1}^{n} \sigma_i^2 \boldsymbol{u}_i \boldsymbol{u}_i^\top$ of the covariance matrix $\boldsymbol{\Sigma} = \sum_{\alpha=1}^{N} \boldsymbol{x}_\alpha \boldsymbol{x}_\alpha^\top$ for the smallest r eigenvalues σ_i^2, $i = 1, ..., r$.

- An rD subspace that best fits the data points $\boldsymbol{x}_\alpha$, $\alpha = 1, ..., N$, is the space spanned by the singular vectors $\boldsymbol{u}_i$, $i = 1, ..., r$, of the singular value decomposition $\boldsymbol{X} = \sum_{i=1}^{n} \sigma \boldsymbol{u}_i \boldsymbol{v}_i^\top$ of the matrix $\boldsymbol{X} = \begin{pmatrix} \boldsymbol{x}_1 & \cdots & \boldsymbol{x}_N \end{pmatrix}$ consisting of the data as its columns for the smallest r singular values σ_i, $i = 1, ..., r$.

- For fitting a subspace, using the singular value decomposition is computationally more efficient than using the spectral decomposition.

- For fitting an affine space, we first compute the centroid of the data points and fit a subspace by regarding it as the origin.

- Using the singular value decomposition for fitting a subspace or an affine space, we can improve the computational efficiency; we need not compute the covariance matrix.

8.6 SUPPLEMENTAL NOTES

As pointed out in Section 8.4, the technique of computing the spectral decomposition of the covariance matrix of Eq. (8.14) and hierarchically fitting rD affine spaces $\mathcal{A}_r$ has been used in various problems of engineering and has been called by many different names. For signal and pattern recognition applications, it is called the *Karhunen–Loéve expansion* (*KL-expansion*). Using this, we can represent signals and patterns with respect to as small a number of basis vectors as possible as long as the residual J can be tolerable and make the transmission and display of the data efficient. This technique is called data compression or image compression.

In statistics, it is called the *principal component analysis*, by which we can grasp the characteristics of multi-dimensional statistical data, extract a small number of statistics that can sufficiently describe them well, and do predictions and tests. These techniques were introduced to computer vision research, and now they are widely used in all types of recognition task. Today, the terms "KL-expansion" and "principal component analysis" are used almost synonymously, and the latter is regarded more or less as generic.

A well-known application is face recognition. An $n \times n$ pixel image is regarded a point in an n^2D space. Given N face images, we can fit to them a lower dimensional subspace after subtracting the centroid, known as the "mean face." The basis vectors of the fitted subspace are often called "eigenfaces." We can classify and recognize faces by ignoring basis vectors with small residual contributions. This technique, often called the "eigenspace method," is also extended to 3D face shapes, i.e., depth images; today, various 3D sensors are available, including the device known as "kinect."

Another application is image surveillance for detecting unusual events. This is done by subtracting from each image frame the background image, but the main difficulty is in background image generation, since it is susceptible to illumination changes due to time of the day and the weather. Also, small fluctuations always exist, including trees in breeze. The eigenspace method is used to extract the underlying structure beneath these fluctuations and generate the "best background" for the current observation.

For geometric modeling of the scene, we describe the camera images and the 3D structures reconstructed from them in terms of lines and planes optimally fitted to them. We can also reduce many video analysis problems to fitting of subspaces and affine spaces in a high-dimensional space. Suppose we track the motion of N points in the scene though m image frames. A image trajectory of one point (x_1, y_1), ..., (x_m, y_m) is regarded as a point in an nD space ($n = 2m$). If we observe the motion of N salient points ("feature points" in the computer vision terminology), we obtain N points in an nD space. From the camera imaging geometry based on perspective projection, many types of 3D analysis of object shapes and camera motions reduce to fitting a subspace or an affine space of a specified dimension to the N points (we will discuss this more in the next chapter). In Section 8.3, we pointed out that using singular value decomposition is much more efficient than using spectral decomposition for $N \ll n$ or $n \ll N$, because the number of operations for singular value decomposition of an $n \times N$ matrix is "linear" in the smaller of n and N (but quadratic in the larger one), while for spectral decomposing it is "cubic" in the size of the matrix.

For actual video processing, accurately extracting feature points by image processing is a difficult task, so the number N of detected feature point is limited. It is also difficult to track them over a long image sequence without disruptions ("occlusion" in the computer vision terminology). As a result, almost all practical applications require either tracking a small number of feature points over a long image sequence or tracking a large number of feature points over a short image sequence. The latter is equivalent to observing a large number of feature points by a small number of cameras. In either case, using singular value decomposition is overwhelmingly faster to compute than using spectral decomposition. This issue is discussed in detail in [12] in relation to 3D reconstruction from multiple views.

8.7 PROBLEMS

8.1. Let λ_1, ..., λ_n be the eigenvalues of an $n \times n$ symmetric matrix $\boldsymbol{A}$. Show the following identity:

$$\mathrm{tr}\boldsymbol{A} = \sum_{i=1}^{n} \lambda_i. \tag{8.22}$$

8.2. Show that the condition for $n+1$ points $\boldsymbol{x}_0, \boldsymbol{x}_1, \ldots, \boldsymbol{x}_n$ in $\mathcal{R}^n$ to be in general position is given by

$$\begin{vmatrix} \boldsymbol{x}_0 & \boldsymbol{x}_1 & \cdots & \boldsymbol{x}_n \\ 1 & 1 & \cdots & 1 \end{vmatrix} \neq 0, \tag{8.23}$$

where the left side is the determinant of an $(n+1) \times (n+1)$ matrix.[1]

8.3. Show that the point $\boldsymbol{g}$ that minimizes the square sum $\sum_{\alpha=1}^{N} \|\boldsymbol{x}_\alpha - \boldsymbol{g}\|^2$ of N points $\{\boldsymbol{x}_\alpha\}$, $\alpha = 1, \ldots, N$, is given by the centroid $\boldsymbol{g}$ given by Eq. (8.13).

8.4. Show that the covariance matrix $\boldsymbol{\Sigma}$ of Eq. (8.14) is also written in the form

$$\boldsymbol{\Sigma} = \sum_{\alpha=1}^{N} \boldsymbol{x}_\alpha \boldsymbol{x}_\alpha^\top - N \boldsymbol{g} \boldsymbol{g}^\top. \tag{8.24}$$

[1]The $(n+1)$D vector $\begin{pmatrix} \boldsymbol{x}_i \\ 1 \end{pmatrix}$ of the ith column expresses the "homogeneous coordinates" of $\boldsymbol{x}_i$ ($\hookrightarrow$ Eq. (7.17)).

CHAPTER 9

Matrix Factorization

In this chapter, we consider the problem of "factorization" of a matrix, i.e., expressing a given matrix $\boldsymbol{A}$ as the product $\boldsymbol{A} = \boldsymbol{A}_1\boldsymbol{A}_2$ of two matrices $\boldsymbol{A}_1$ and $\boldsymbol{A}_2$. We discuss its relationship to the matrix rank and the singular value decomposition. As a typical application, we describe a technique, called the "factorization method," for reconstructing the 3D structure of the scene from images captured by multiple cameras.

9.1 MATRIX FACTORIZATION

Suppose we want to express an $m \times n$ matrix $\boldsymbol{A}$ as the product of two matrices $\boldsymbol{A}_1$ and $\boldsymbol{A}_2$ in the form

$$\boldsymbol{A} = \boldsymbol{A}_1\boldsymbol{A}_2, \tag{9.1}$$

where $\boldsymbol{A}_1$ and $\boldsymbol{A}_2$ are $m \times r$ and $r \times n$ matrices, respectively. We assume $r \leq m, n$. We call this problem matrix *factorization*. When such a problem appears in engineering applications, usually some properties are required for $\boldsymbol{A}_1$ and $\boldsymbol{A}_2$ to satisfy.

Evidently, the decomposition of the form of Eq. (9.1) is not unique. In fact, if such matrices $\boldsymbol{A}_1$ and $\boldsymbol{A}_2$ are obtained, the matrices

$$\boldsymbol{A}_1' = \boldsymbol{A}_1\boldsymbol{C}^{-1}, \qquad \boldsymbol{A}_2' = \boldsymbol{C}\boldsymbol{A}_2 \tag{9.2}$$

for an arbitrary $r \times r$ nonsingular matrix $\boldsymbol{C}$ satisfy $\boldsymbol{A}_1\boldsymbol{A}_2 = \boldsymbol{A}_1'\boldsymbol{A}_2'$. In a real application, we first tentatively compute some matrices $\boldsymbol{A}_1$ and $\boldsymbol{A}_2$ that satisfy Eq. (9.1) and then find the nonsingular matrix $\boldsymbol{C}$ in such a way that the required properties imposed on matrices $\boldsymbol{A}_1'$ and $\boldsymbol{A}_2'$ of Eq. (9.2) are satisfied.

If no special dependencies exist among columns and rows of a matrix, its rank generally coincides with the smaller of the numbers of columns and rows. Assume that matrices $\boldsymbol{A}_1$ and $\boldsymbol{A}_2$ both have rank r. It is known that the rank of the product of two matrices does not exceed the rank of either one. Namely,

$$\mathrm{rank}(\boldsymbol{A}\boldsymbol{B}) \leq \min(\mathrm{rank}(\boldsymbol{A}), \mathrm{rank}(\boldsymbol{B})) \tag{9.3}$$

for any matrices $\boldsymbol{A}$ and $\boldsymbol{B}$ for which their product can be defined.

Equation (9.3) can be explained as follows (Fig. 9.1). Let $\boldsymbol{A}$ and $\boldsymbol{B}$ be $l \times m$ and $m \times n$ matrices, respectively. An $l \times m$ matrix $\boldsymbol{A}$ defines a linear mapping from $\mathcal{R}^m$ to $\mathcal{R}^l$. We write $\boldsymbol{A}(\mathcal{R}^m)$ for the image of $\mathcal{R}^m$ by $\boldsymbol{A}$, i.e., the subspace spanned by the vectors obtained by mapping the basis of $\mathcal{R}^m$ by $\boldsymbol{A}$. The dimension of $\boldsymbol{A}(\mathcal{R}^m)$ (= the number of independent columns

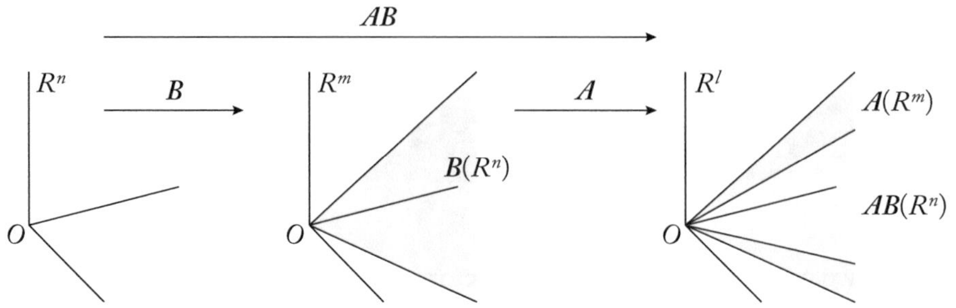

Figure 9.1: Composition of linear mappings. The image $\boldsymbol{A}\boldsymbol{B}(\mathcal{R}^n)$ of $\mathcal{R}^n$ by the product mapping $\boldsymbol{A}\boldsymbol{B}$ equals the mapping of $\mathcal{R}^n$ by $\boldsymbol{B}$ followed by the mapping $\boldsymbol{A}$.

of $\boldsymbol{A}$) is $\text{rank}(\boldsymbol{A})$. Similarly, an $m \times n$ matrix $\boldsymbol{B}$ defines a linear mapping from $\mathcal{R}^n$ to $\mathcal{R}^m$, and the dimension of $\boldsymbol{B}(\mathcal{R}^n)$ is $\text{rank}(\boldsymbol{B})$. On the other hand, since $(\boldsymbol{A}\boldsymbol{B})(\mathcal{R}^n)$ is obtained by first mapping $\mathcal{R}^n$ by $\boldsymbol{B}$ and then mapping it by $\boldsymbol{A}$, it is a subset of $\boldsymbol{A}(\mathcal{R}^m)$. Hence, the dimension of $(\boldsymbol{A}\boldsymbol{B})(\mathcal{R}^n)$ ($= \text{rank}(\boldsymbol{A}\boldsymbol{B})$) does not exceed the dimension of $\boldsymbol{A}(\mathcal{R}^m)$ ($= \text{rank}(\boldsymbol{A})$). Thus, we obtain $\text{rank}(\boldsymbol{A}\boldsymbol{B}) \leq \text{rank}(\boldsymbol{A})$. Using the same argument, we obtain $\text{rank}(\boldsymbol{B}^\top \boldsymbol{A}^\top) \leq \text{rank}(\boldsymbol{B}^\top)$. However, the number of linearly independent columns of a matrix equals the number of its linearly independent rows. Hence, $\text{rank}(\boldsymbol{B}^\top) = \text{rank}(\boldsymbol{B})$ and $\text{rank}(\boldsymbol{B}^\top \boldsymbol{A}^\top) = \text{rank}(\boldsymbol{A}\boldsymbol{B})$. Consequently, we also obtain $\text{rank}(\boldsymbol{A}\boldsymbol{B}) \leq \text{rank}(\boldsymbol{B})$. □

From this observation, we find that for computing the factorization of Eq. (9.1), the $m \times n$ matrix $\boldsymbol{A}$ must have rank r or less. However, if $\boldsymbol{A}$ involves measurement data, its rank is generally equal to either m or n (the smaller one). Hence, if $r < m, n$, the decomposition of Eq. (9.1) is not possible. In such a case, we compute such $\boldsymbol{A}_1$ and $\boldsymbol{A}_2$ that the decomposition of Eq. (9.1) approximately holds (↪ Problem 9.1). This is done by minimally modifying $\boldsymbol{A}$ so that it has rank r. To be specific, we constrain the rank as discussed in Section 5.4 and replace $\boldsymbol{A}$ by $(\boldsymbol{A})_r$ (↪ Footnote 1 of Section 5.5).

Then, we can determine $\boldsymbol{A}_1$ and $\boldsymbol{A}_2$ that satisfy $(\boldsymbol{A})_r = \boldsymbol{A}_1\boldsymbol{A}_2$; the decomposition is not unique. A simple way to obtain a candidate solution is to first compute the singular value decomposition

$$(\boldsymbol{A})_r = \boldsymbol{U}\boldsymbol{\Sigma}\boldsymbol{V}^\top,$$

$$\boldsymbol{U} = \begin{pmatrix} \boldsymbol{u}_1 & \cdots & \boldsymbol{u}_r \end{pmatrix}, \qquad \boldsymbol{\Sigma} = \begin{pmatrix} \sigma_1 & & \\ & \ddots & \\ & & \sigma_r \end{pmatrix}, \qquad \boldsymbol{V} = \begin{pmatrix} \boldsymbol{v}_1 & \cdots & \boldsymbol{v}_r \end{pmatrix}, \tag{9.4}$$

as in Eq. (4.10), and then factorize the diagonal matrix $\boldsymbol{\Sigma}$ into the form $\boldsymbol{\Sigma} = \boldsymbol{\Sigma}_1\boldsymbol{\Sigma}_2$. Finally, we obtain $\boldsymbol{A}_1$ and $\boldsymbol{A}_2$ in the form

$$\boldsymbol{A}_1 = \boldsymbol{U}\boldsymbol{\Sigma}_1, \qquad \boldsymbol{A}_2 = \boldsymbol{\Sigma}_2\boldsymbol{V}. \tag{9.5}$$

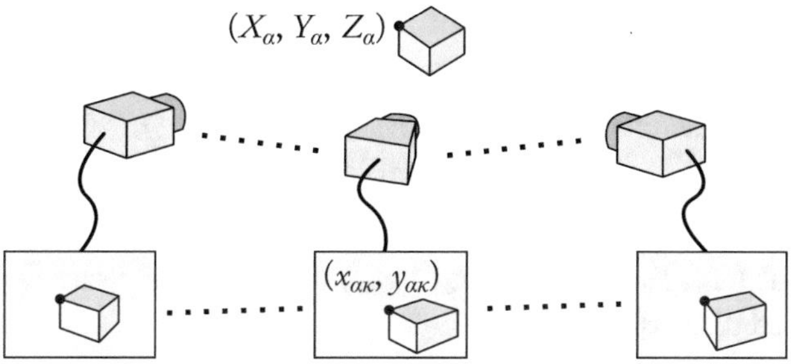

Figure 9.2: N points $(X_\alpha, Y_\alpha, Z_\alpha)$, $\alpha = 1, ..., N$ in the 3D scene are captured by M cameras. The αth point is imaged at $(x_{\alpha\kappa}, y_{\alpha\kappa})$, $\kappa = 1, ..., M$, in the image of the κth camera.

Typical factorizations of $\mathbf{\Sigma}$ are

$$\text{(i)} \quad \mathbf{\Sigma}_1 = \begin{pmatrix} \sigma_1 & & \\ & \ddots & \\ & & \sigma_r \end{pmatrix}, \qquad \mathbf{\Sigma}_2 = \boldsymbol{I}, \tag{9.6}$$

$$\text{(ii)} \quad \mathbf{\Sigma}_1 = \begin{pmatrix} \sqrt{\sigma_1} & & \\ & \ddots & \\ & & \sqrt{\sigma_r} \end{pmatrix}, \qquad \mathbf{\Sigma}_2 = \begin{pmatrix} \sqrt{\sigma_1} & & \\ & \ddots & \\ & & \sqrt{\sigma_r} \end{pmatrix}, \tag{9.7}$$

$$\text{(iii)} \quad \mathbf{\Sigma}_1 = \boldsymbol{I}, \qquad \mathbf{\Sigma}_2 = \begin{pmatrix} \sigma_1 & & \\ & \ddots & \\ & & \sigma_r \end{pmatrix}. \tag{9.8}$$

The diagonal matrix $\mathbf{\Sigma}_1 = \mathbf{\Sigma}_2$ in (ii) is also written as $\sqrt{\mathbf{\Sigma}}$. From this factorization procedure, we also see that the condition for an $m \times n$ matrix $\boldsymbol{A}$ to have rank r or less ($r \leq m, n$) is that it can be written as $\boldsymbol{A} = \boldsymbol{A}_1\boldsymbol{A}_2$ for some $m \times r$ matrix $\boldsymbol{A}_1$ and some $r \times n$ matrix $\boldsymbol{A}_2$.

9.2 FACTORIZATION FOR MOTION IMAGE ANALYSIS

Suppose we take images of N points $(X_\alpha, Y_\alpha, Z_\alpha)$, $\alpha = 1, ..., N$, in the 3D scene, using M cameras (or equivalently moving one camera). Suppose the αth point is projected to $(x_{\alpha\kappa}, y_{\alpha\kappa})$ in the image plane of the κth camera (Fig. 9.2).

We define an XYZ coordinate system in the scene so that its origin O coincides with the centroid of the N points:

$$\sum_{\alpha=1}^{N} X_\alpha = \sum_{\alpha=1}^{N} Y_\alpha = \sum_{\alpha=1}^{N} Z_\alpha = 0. \tag{9.9}$$

We assume that all the points are seen by all the cameras and we define an image coordinate system in each image so that its origin (0, 0) coincides with the centroid of the N image positions:

$$\sum_{\alpha=1}^{N} x_{\alpha\kappa} = \sum_{\alpha=1}^{N} y_{\alpha\kappa} = 0, \qquad \kappa = 1, ..., M. \tag{9.10}$$

Then, it is known that $(X_\alpha, Y_\alpha, Z_\alpha)$ and $(x_{\alpha\kappa}, y_{\alpha\kappa})$ approximately satisfy the relationship (see Section 9.3 for detailed discussions)

$$\begin{pmatrix} x_{\alpha\kappa} \\ y_{\alpha\kappa} \end{pmatrix} = \boldsymbol{\Pi}_\kappa \begin{pmatrix} X_\alpha \\ Y_\alpha \\ Z_\alpha \end{pmatrix}, \tag{9.11}$$

where $\boldsymbol{\Pi}_\kappa$ is a 2×3 matrix, called the *camera matrix*, determined by the position and orientation of the κth camera and its internal parameters.

Strictly speaking, the camera imaging geometry is described by a nonlinear relationship, called *perspective projection* (↪ Fig. 7.2). If we ignore the perspective effect, which causes objects in the distance to look small, we obtain the linear approximation of Eq. (9.11) (we will discuss this more in Section 9.3). It is known that this approximation holds well when the objects we are viewing are relatively in the distance, e.g., persons standing several meters away, or small objects are zoomed in within a relatively small region of the image. Hypothetical cameras for which Eq. (9.11) holds are said to be *affine*.

Arrange all observed points $(x_{\alpha\kappa}, y_{\alpha\kappa})$, $\kappa = 1, ..., M$, $\alpha = 1, ..., N$, in all images in an $2M \times N$ matrix in the form

$$\boldsymbol{W} = \begin{pmatrix} x_{11} & \cdots & x_{N1} \\ y_{11} & \cdots & y_{N1} \\ \vdots & \ddots & \vdots \\ x_{1M} & \cdots & x_{NM} \\ y_{1M} & \cdots & y_{NM} \end{pmatrix}, \tag{9.12}$$

which we call the *observation matrix*. Arrange all camera matrices $\boldsymbol{\Pi}_\kappa$, $\kappa = 1, ..., M$, and all 3D coordinates $(X_\alpha, Y_\alpha, Z_\alpha)$, $\alpha = 1, ..., N$, in the matrix form

$$\boldsymbol{M} = \begin{pmatrix} \boldsymbol{\Pi}_1 \\ \vdots \\ \boldsymbol{\Pi}_M \end{pmatrix}, \qquad \boldsymbol{S} = \begin{pmatrix} X_1 & \cdots & X_N \\ Y_1 & \cdots & Y_N \\ Z_1 & \cdots & Z_N \end{pmatrix}. \tag{9.13}$$

We call the $2M \times 3$ matrix $\boldsymbol{M}$ the *motion matrix* and the $3 \times N$ matrix $\boldsymbol{N}$ the *shape matrix*.

From the definition of the matrix $\boldsymbol{W}$ of Eq. (9.11) and the definition of the matrices $\boldsymbol{M}$ and $\boldsymbol{N}$ of Eq. (9.13), we obtain the following relationship (↪ Problem 9.2):

$$\boldsymbol{W} = \boldsymbol{M}\boldsymbol{S}. \tag{9.14}$$

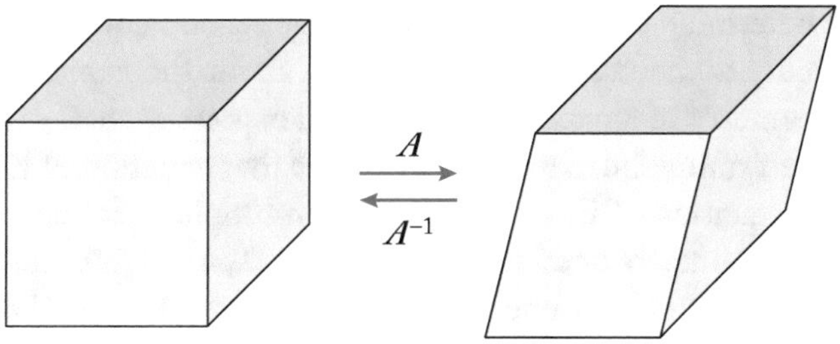

Figure 9.3: Affine reconstruction. The computed shape and its true shape are related by an unknown affine transformation A. The lengths and angles may alter, but the ratio of lengths is preserved, and parallel lines and planes are mapped to parallel lines and planes.

It follows that if the matrix W obtained from the coordinates of the points observed in images is decomposed into the product of the matrices M and S by the method described in the preceding section, all the camera matrices and all the 3D point positions are obtained. This technique of reconstructing the 3D shape from images is called the *factorization method*.

However, the solution is not unique, as pointed out in the preceding section. If $\bar{M}$ and $\bar{S}$ are the true motion matrix and the shape matrix, respectively, the matrices M and S obtained by the factorization method are related to $\bar{M}$ and $\bar{S}$ by

$$M = \bar{M}A^{-1}, \qquad S = A\bar{S}, \tag{9.15}$$

for some 3×3 nonsingular matrix A. The second equation states that each column $(X_\alpha, Y_\alpha, Z_\alpha)^\top$ of S equals multiplication of each column $(\bar{X}_\alpha, \bar{Y}_\alpha, \bar{Z}_\alpha)^\top$ of $\bar{S}$ by some nonsingular matrix A. It follows that the computed 3D shape is a linear transformation of the true shape. Since the absolute position is indeterminate,[1] it is an *affine transformation* of the true shape.

Affine transformations preserve collinearity and coplanarity (Fig. 9.3), i.e., collinear points are mapped to collinear points, and coplanar points are mapped to coplanar points. As a result, parallel lines and planes are mapped to parallel lines and planes. However, scales and angles may change. For example, a cube is mapped to a parallelepiped. 3D reconstruction up to indeterminacy of affine transformations is called an *affine reconstruction*.

In order to remove this indeterminacy and obtain a shape with correct angles, which we call a *Euclidean reconstruction*,[2] we need to specify the indeterminate matrix A, using some knowledge or constraint, which we call the *metric condition*. One possibility is the use of the knowledge about the 3D scene. For example, we require a particular edge to be orthogonal to

[1] Since the origin of the coordinate system in the scene is defined by Eq. (8.7) for the sake of computational convenience, the absolute position in the scene is indeterminate.

[2] Since the absolute scale is indeterminate from images alone (↪ Section 7.3), we should call it a "similar" reconstruction to be strict, but this term is widely used.

a particular edge and determine the matrix $\boldsymbol{A}$ so that the second equation of Eq. (9.15) is satisfied. Another possibility is the use of the knowledge about the camera. For this, we model the camera imaging geometry in a parametric form and express each $\boldsymbol{\Pi}_\kappa$ in terms of unknown parameters. Then, we determine the matrix $\boldsymbol{A}$ so that the first equation of Eq. (9.15) is satisfied. To be specific, the first equation of Eq. (9.15) leads to multiple equalities, from which we can eliminate the unknown parameters of each camera matrix $\boldsymbol{\Pi}_\kappa$ to obtain the unknown matrix $\boldsymbol{A}$. To this end, various parametric affine camera models are proposed. Typical affine camera models include *orthographic projection*, *weak perspective projection*, and *paraperspective projection* (we will discuss these in Section 9.3).

GLOSSARY AND SUMMARY

affine camera: A camera for which the imaging geometry is described by linear equations that approximate the perspective projection. Such a camera is merely a mathematical model and does not exist in reality.

affine reconstruction: A 3D reconstruction method such that the computed shape and its true shape are related by an unknown affine transformation.

affine transformation: A mapping defined by a linear transformation followed by a translation. Collinear and coplanar points are mapped to collinear and coplanar points, but scales and angles change, so that a cube is mapped to a parallelepiped, for example.

camera matrix: The matrix that describes the relationship between the positions of points in 3D and their 2D coordinates in the captured images.

Euclidean reconstruction: A 3D reconstruction method such that, as opposed to affine reconstruction, the computed shape has correct angles. Mathematically speaking, this should be called "similar reconstruction," but this term is widely used.

factorization: Expressing a matrix $\boldsymbol{A}$ as the product $\boldsymbol{A} = \boldsymbol{A}_1\boldsymbol{A}_2$ of two matrices $\boldsymbol{A}_1$ and $\boldsymbol{A}_2$.

factorization method: The technique for computing the 3D positions of multiple points and the positions, orientations, and internal parameters of multiple cameras from their images by assuming that the cameras are affine. The computation reduces to matrix factorization.

metric condition: Conditions required to transform an affine reconstruction to a Euclidean reconstruction.

motion matrix: The matrix consisting of the positions, orientations, and internal parameters of the cameras that capture multiple points in 3D.

observation matrix: The matrix consisting of image coordinates of multiple points in 3D that are captured by multiple cameras.

perspective projection: The camera imaging is modeled as the intersections of the image plane with lines of sight passing through the lens center. Ordinary cameras are well described by this model. As the object we are viewing goes away in the distance, its imaged size becomes smaller.

shape matrix: The matrix consisting of 3D coordinates of multiple points that are imaged.

- The expression of a given matrix as the product of two matrices of specified sizes is not unique; it has indeterminacy of an arbitrary nonsingular matrix (Eq. (9.2)).
- The rank of a matrix given by the product of two matrices is not more than the rank of each matrix (Eq. (9.3)). The matrix that does not satisfy this condition does not allow such a factorization.
- If we want to approximately factorize a matrix that does not satisfy the rank requirement into the product of two matrices of specified ranks, we first compute the singular value decomposition and adjust the rank by truncating small singular values.
- Factorization of a matrix that satisfies the rank constraint reduces to factorization of diagonal matrices via the singular value decomposition.
- Given image coordinates of multiple points in 3D that are captured in images taken by multiple cameras, we can compute the camera positions and 3D shapes by factorizing the observation matrix consisting of image coordinates, assuming that the cameras are affine. However, the obtained result is merely an affine reconstruction.
- For correcting the affine reconstruction obtained by the factorization method to a Euclidean reconstruction, we assume some knowledge of the 3D scene. Alternatively, we introduce specific camera modeling, parameterize the cameras, and eliminate the parameters from resulting multiple equations.

9.3 SUPPLEMENTAL NOTES

As mentioned in Section 7.2, the standard camera imaging is modeled as *perspective projection*: a point $P : (X, Y, Z)$ is imaged at the intersection of the ray OP starting from the *viewpoint* O (the lens center) with an image plane orthogonal to the *optical axis* (the axis of symmetry of the lens) (Fig. 7.2). The distance f from the viewpoint O to the imaging plane is called the *focal length*.

If the αth point $(X_\alpha, Y_\alpha, Z_\alpha)$ in the 3D scene is captured by the κth camera on its image plane at $(x_{\alpha\kappa}, y_{\alpha\kappa})$, the imaging relationship under the standard perspective camera model is given, instead of Eq. (9.11), in terms of *homogeneous coordinates* (we discuss this shortly) in the form

$$\begin{pmatrix} x_{\alpha\kappa}/f_0 \\ y_{\alpha\kappa}/f_0 \\ 1 \end{pmatrix} \simeq \boldsymbol{P}_\kappa \begin{pmatrix} X_\alpha \\ Y_\alpha \\ Z_\alpha \\ 1 \end{pmatrix}, \tag{9.16}$$

for some 3×4 matrix $\boldsymbol{P}_\kappa$ [5, 12], where $\simeq$ denotes equality up to an unknown scale and f_0 is an appropriate scaling constant to stabilize finite length numerical computation [2]. The matrix $\boldsymbol{P}_\kappa$, called the *camera matrix* of the κth (perspective) camera, consists of the position, orientation, and internal parameters, including the focal length, of the κth camera. *Affine camera modeling* means approximating Eq. (9.16) by

$$\begin{pmatrix} x_{\alpha\kappa}/f_0 \\ y_{\alpha\kappa}/f_0 \\ 1 \end{pmatrix} = \boldsymbol{P}_\kappa \begin{pmatrix} X_\alpha \\ Y_\alpha \\ Z_\alpha \\ 1 \end{pmatrix}. \tag{9.17}$$

The difference is that the relation symbol $\simeq$ is replaced by the equality $=$. As a result, Eq. (9.17) no longer describes perspective projection, but it is known to be a good approximation if the scene is in the distance or the portion we want to reconstruct is imaged around the *principal point* (the intersection of the optical axis with the image plane) within a small region as compared with the focal length [5, 12].

If all points are observed in all images, Eq. (9.17) is simplified as follows. We choose the world coordinate origin to be the centroid of the points $(X_\alpha, Y_\alpha, Z_\alpha)$ so that Eq. (9.9) holds and choose the image coordinate origin to be the centroid of the observed positions $(x_{\alpha\kappa}, y_{\alpha\kappa})$ for each camera so that Eq. (9.10) holds. Hence, operation $(1/N)\sum_{\alpha=1}^{N}$ on both sides of Eq. (9.17) leads to $(0, 0, 1)^\top = \boldsymbol{P}_\kappa(0, 0, 0, 1)^\top$. This means that Eq. (9.17) has the form

$$\begin{pmatrix} x_{\alpha\kappa}/f_0 \\ y_{\alpha\kappa}/f_0 \\ 1 \end{pmatrix} = \begin{pmatrix} & \boldsymbol{\Pi}_\kappa & & \begin{matrix} 0 \\ 0 \end{matrix} \\ 0 & 0 & 0 & 1 \end{pmatrix} \begin{pmatrix} X_\alpha \\ Y_\alpha \\ Z_\alpha \\ 1 \end{pmatrix}, \tag{9.18}$$

where $\boldsymbol{\Pi}_\kappa$ is some 2×3 matrix. This is further simplified to Eq. (9.11), where the scaling constant f_0 is omitted. Since its role is merely to make the components of the vector on the left side of Eqs. (9.16) and (9.17) have the same order of magnitude, we can let $f_0 = 1$.

As stated in Section 9.2, the 3D scene reconstructed by the factorization method based on affine camera modeling is merely affine reconstruction up to an unknown affine transformation. In order to make it Euclidean reconstruction using the knowledge of the cameras, we need to give an appropriate parametric form of the camera matrix $\boldsymbol{\Pi}_\kappa$. Typical parametric models include *orthographic projection*, *weak perspective projection*, and *paraperspective projection*. The

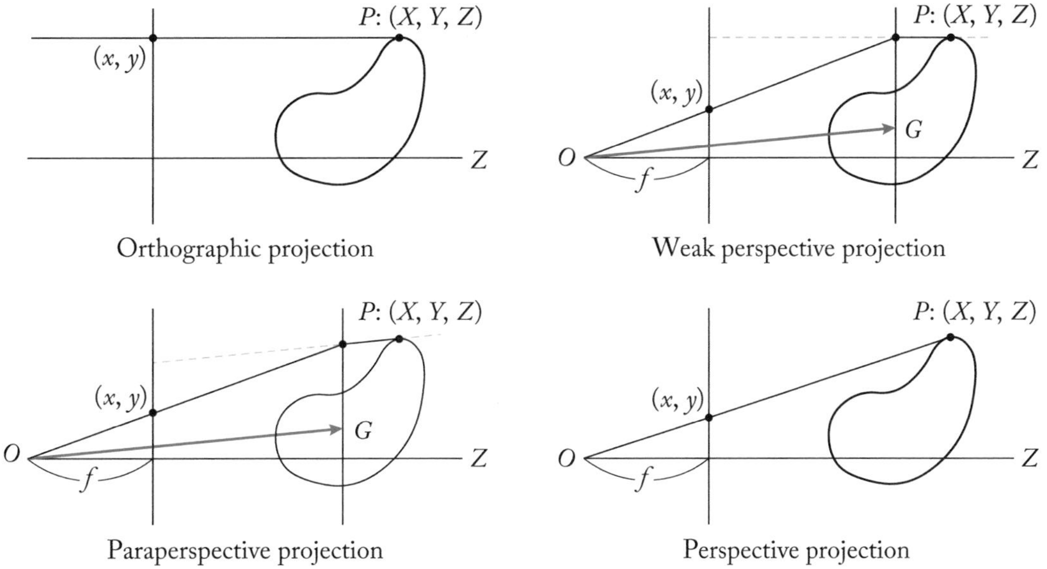

Figure 9.4: Affine camera models (top three) that approximate perspective projection (bottom). The origin O corresponds to the lens center (infinitely far away for orthographic cameras) and the Z-axis is the optical axis. The point G indicates the object centroid. The lay starting from O is parallel to the optical axis OZ for orthographic projection. For weak perspective and paraperspective projections, it first reaches the plane orthogonal to OZ and passing through G. Then, it goes to P in parallel to OZ for weak perspective projection and in parallel to OG for paraperspective projection. All these approximate the ray OP for perspective projection.

imaging geometry of these models are schematically illustrated in Fig. 9.4, where the origin O corresponds to the lens center (infinitely far away for orthographic cameras) and the Z-axis is the optical axis. The point G indicates the object centroid.

The factorization method was first presented by Tomasi and Kanade [21] for orthographic cameras. The technique was extended to weak perspective cameras and paraperspective cameras by Poelman and Kanade [17]. They showed that the affine mapping $\boldsymbol{A}$ in Eq. (9.15) for fixing a Euclidean reconstruction is obtained by computing the 3×3 symmetric matrix

$$\boldsymbol{T} = \boldsymbol{A}\boldsymbol{A}^\top. \tag{9.19}$$

However, these popular affine camera models are mutually unrelated without any hierarchical relations. Namely, orthographic projection is not a special case of weak perspective or paraperspective projection, and weak perspective projection is not a special case of paraperspective projection. Later, Kanatani et al. [10] proposed what they called the *symmetric affine camera* model and pointed out that orthographic, weak perspective, and paraperspective projections are all special cases of the symmetric affine camera model. They showed that Euclidean reconstruction is possible and the matrix $\boldsymbol{T}$ of Eq. (9.19) can be determined only from that generic modeling (see [12] for the computational procedure).

The beauty of the mathematical formulation of 3D reconstruction by factorization based on affine cameras motivated many researchers to extend it to perspective cameras, and a sophisticated mathematical framework has been established based on projective geometry. Here, we summarize its outline.

The first step is to represent a point in 3D by *homogeneous coordinates* $X_1 : X_2 : X_3 : X_4$. They are related to the usual coordinates (X, Y, Z), called *inhomogeneous coordinates*, by

$$X = \frac{X_1}{X_4}, \qquad Y = \frac{X_2}{X_4}, \qquad Z = \frac{X_3}{X_4}. \tag{9.20}$$

Namely, the homogeneous coordinates are defined only up to scale, and only the ratios among them have meaning. Points for which $X_4 = 0$ are interpreted to be at infinitely far away. Then, Eq. (9.16) is written as

$$z_{\alpha\kappa} \begin{pmatrix} x_{\alpha\kappa}/f_0 \\ y_{\alpha\kappa}/f_0 \\ 1 \end{pmatrix} = \boldsymbol{P}_\kappa \boldsymbol{X}_\alpha, \tag{9.21}$$

where $\boldsymbol{X}_\alpha$ is a 4D vector consisting of the homogeneous coordinates and $z_{\alpha\kappa}$ is an unknown proportionality constant, called the *projective depth*.

We write all the observed image points $(x_{\alpha\kappa}, y_{\alpha\kappa})$ and their projective depths $z_{\alpha\kappa}$, $\alpha = 1$, ..., N, $\kappa = 1$, ..., M, in the following matrix form:

$$\boldsymbol{W} = \begin{pmatrix} z_{11}x_{11}/f_0 & z_{21}x_{21}/f_0 & \cdots & z_{N1}x_{N1}/f_0 \\ z_{11}y_{11}/f_0 & z_{21}y_{21}/f_0 & \cdots & z_{N1}y_{N1}/f_0 \\ z_{11} & z_{21} & \cdots & z_{N1} \\ \vdots & \vdots & \ddots & \vdots \\ z_{1M}x_{1M}/f_0 & z_{2M}x_{2M}/f_0 & \cdots & z_{NM}x_{NM}/f_0 \\ z_{1M}y_{1M}/f_0 & z_{2M}y_{2M}/f_0 & \cdots & z_{NM}y_{NM}/f_0 \\ z_{1M} & z_{2M} & \cdots & z_{NM} \end{pmatrix}. \tag{9.22}$$

Adopting the same terminology as in the case of affine cameras, we call this $3M \times N$ matrix the *observation matrix*. We also arrange the matrices $\boldsymbol{P}_\kappa$ of all the cameras and the homogeneous coordinate vectors $\boldsymbol{X}_\alpha$ of all the points as matrices in the form

$$\boldsymbol{M} = \begin{pmatrix} \boldsymbol{P}_1 \\ \vdots \\ \boldsymbol{P}_M \end{pmatrix}, \qquad \boldsymbol{S} = \begin{pmatrix} \boldsymbol{X}_1 & \cdots & \boldsymbol{X}_N \end{pmatrix}. \tag{9.23}$$

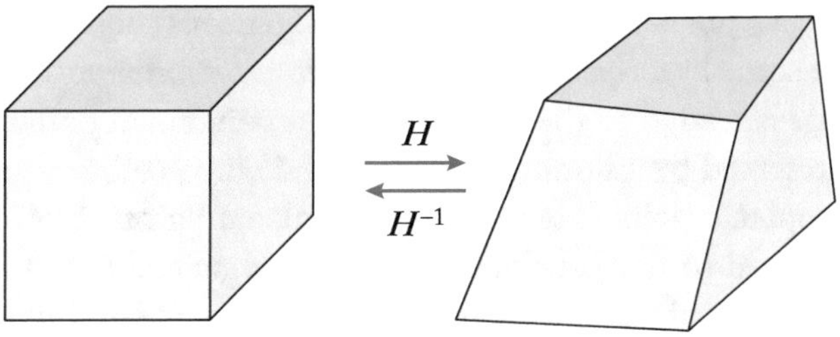

Figure 9.5: Projective reconstruction. The reconstructed 3D shape is related to the true shape by an unknown homography $\boldsymbol{H}$. The lengths, angles, and ratios may alter, but lines and planes are mapped to lines and planes.

As in the case of affine reconstruction, we call the $3M \times 4$ matrix $\boldsymbol{M}$ and the $4 \times N$ matrix $\boldsymbol{S}$ the *motion matrix* and the *shape matrix*, respectively. From Eq. (9.21) and the above definition of the matrices $\boldsymbol{S}$ and $\boldsymbol{M}$, we obtain

$$\boldsymbol{W} = \boldsymbol{M}\boldsymbol{S}. \tag{9.24}$$

Hence, if the projective depths $z_{\alpha\kappa}$ are known, we can determine $\boldsymbol{M}$ and $\boldsymbol{S}$ by the factorization of $\boldsymbol{W}$ and obtain the camera matrices $\boldsymbol{P}_\kappa$ and the 3D positions $\boldsymbol{X}_\alpha$. Thus, the task reduces to find the projective depths $z_{\alpha\kappa}$ such that the matrix $\boldsymbol{W}$ of Eq. (9.22) can be factored into the product of some $3M \times 4$ matrix $\boldsymbol{M}$ and some $4 \times N$ matrix $\boldsymbol{S}$. The matrix $\boldsymbol{W}$ is factored in this way if and only if $\boldsymbol{W}$ has rank 4. The rank of a matrix is the number of independent columns or independent rows of it. Hence, we can take either of the following two approaches.

Primary method: We determine $z_{\alpha\kappa}$ so that the N columns of $\boldsymbol{W}$ of Eq. (9.22) span a 4D subspace.

Dual method: We determine $z_{\alpha\kappa}$ so that the $3M$ rows of $\boldsymbol{W}$ of Eq. (9.22) span a 4D subspace.

The primary method was formulated by Mahamud and Hebert [16]. This is done by iterations. First, we let $z_{\alpha\kappa} = 1$. This is equivalent to assuming affine cameras. Then, we fit a 4D subspace to the N columns of $\boldsymbol{W}$ using the singular value decomposition as described in Section 8.3. Next, we modify all $z_{\alpha\kappa}$ so that all the columns of $\boldsymbol{W}$ come closer to the fitted 4D subspace. Then, we newly fit a 4D subspace and iterate this process until all the columns are sufficiently close to the fitted 4D subspace. The dual method was formulated by Heyden et al. [6]. Instead of columns, they iteratively fitted a 4D subspace to the $3M$ rows of $\boldsymbol{W}$ and repeated this until all the rows are sufficiently close to the fitted 4D subspace.

As in the case of affine reconstruction, the factorization of $\boldsymbol{W}$ in the form of Eq. (9.24) is not unique. In fact, if $\boldsymbol{P}_\kappa$ and $\boldsymbol{X}_\alpha$ are multiplied by an arbitrary 4×4 nonsingular matrix $\boldsymbol{H}$ in the form

$$\boldsymbol{P}'_\kappa = \boldsymbol{P}_\kappa \boldsymbol{H}^{-1}, \qquad \boldsymbol{X}'_\alpha = \boldsymbol{H}\boldsymbol{X}_\alpha. \tag{9.25}$$

the relation $\boldsymbol{P}'_\kappa \boldsymbol{X}'_\alpha = \boldsymbol{P}_\kappa \boldsymbol{X}_\alpha$ holds. Multiplying the homogeneous coordinate vector $\boldsymbol{X}_\alpha$ by matrix $\boldsymbol{H}$ means mapping the scene by a *homography*, or a *projective transformation*. This means that the reconstructed 3D shape is related to the true shape by an unknown homography $\boldsymbol{H}$. Linearity and coplanarity are preserved by a homography (Fig. 2.1), i.e., collinear points are mapped to collinear points, and coplanar points are mapped to coplanar points. However, lengths, angles, and ratios are not preserved so that a cube is mapped to a general hexahedron and a sphere is mapped to an ellipsoid. Thus, the reconstructed 3D shape has indeterminacy of a homography and is called *projective reconstruction*.

In order to transform, or "upgrade," it to Euclidean reconstruction, we need to determine the homography matrix $\boldsymbol{H}$ in Eq. (9.25). Triggs [22] showed that corresponding to Eq. (9.19) for affine cameras, determination of $\boldsymbol{H}$ reduces to computation of the 4×4 symmetric matrix

$$\boldsymbol{\Omega} = \boldsymbol{H}\,\mathrm{diag}(1, 1, 1, 0)\boldsymbol{H}^\top, \tag{9.26}$$

which he called the *dual absolute quadric*, or *DAQ* for short, giving it a geometric interpretation in terms of projective geometry. The DAQ can be determined if we have some knowledge of the camera. Seo and Heyden [19] proposed an iterative procedure for determining $\boldsymbol{\Omega}$ based on the knowledge that the photocell array has no skew and the aspect ratio is 1. Later, the method was improved by Kanatani [9]. See [12] for the detailed computational procedures.

9.4 PROBLEMS

9.1. Show that an $m \times n$ matrix $\boldsymbol{A}$ has rank r or less ($r \leq m, n$) if and only if it is written as $\boldsymbol{A} = \boldsymbol{A}_1\boldsymbol{A}_2$ for some $m \times r$ matrix $\boldsymbol{A}_1$ and some $r \times n$ matrix $\boldsymbol{A}_2$.

9.2. (1) The αth column of Eq. (9.12) lists the x- and y-coordinates of the αth point over the M images, which can be seen as the "trajectory" of the αth point. In other words, the trajectory of each point is a point in a $2M$D space. Show that Eq. (9.14) implies that the N points that represent the trajectories in the $2M$D space are all included in a 3D subspace.

(2) Show how to compute an orthonormal basis of that 3D subspace, taking into consideration that the decomposition of Eq. (9.14) is for hypothetical cameras, i.e., affine cameras, and that Eq. (9.14) does not exactly hold for the observation matrix $\boldsymbol{W}$ obtained from real cameras.

CHAPTER 10

Triangulation from Three Views

In this chapter, we consider the problem of reconstructing 3D points from three images. The basic principle is to optimally correct observed image points in the presence of noise, in such a way that their rays, or their lines of sight, intersect at a single point in the scene. We show that three rays intersect in the scene if and only if the "trilinear constraint," specified by the "trifocal tensor," is satisfied. We discuss its geometric meaning and point out the role of pseudoinverse in the course of iterative optimal computation.

10.1 TRINOCULAR STEREO VISION

Suppose we view a point (X, Y, Z) in the 3D scene by three cameras, observing three points (x_0, y_0), (x_1, y_1), and (x_2, y_2) in their camera images. We want to compute (X, Y, Z) from them. This task is called *triangulation* from three views, and the camera setting for this is called *trinocular stereo vision* (Fig. 10.1).

We write the three points (x_0, y_0), (x_1, y_1), and (x_2, y_2) as 3D vectors in the form

$$\boldsymbol{x}_0 = \begin{pmatrix} x_0/f_0 \\ y_0/f_0 \\ 1 \end{pmatrix}, \qquad \boldsymbol{x}_1 = \begin{pmatrix} x_1/f_0 \\ y_1/f_0 \\ 1 \end{pmatrix}, \qquad \boldsymbol{x}_2 = \begin{pmatrix} x_2/f_0 \\ y_2/f_0 \\ 1 \end{pmatrix}, \tag{10.1}$$

where f_0 is a fixed scaling constant. Describing a 2D point by a 3D vector by padding the coordinates with 1 corresponds to the use of *homogeneous coordinates*, the standard convention in projective geometry ($\hookrightarrow$ Eq. (7.17), Footnote 1 of Section 8.7, and Eqs. (9.16)–(9.18)). The constant f_0 is for appropriately scaling the x and y components. Mathematically, it is arbitrary and can be set 1. However, it is known that finite length numerical computation becomes stable if vector components have the same order of magnitude [2]; usually, it is taken to be approximately the size of the image frame.

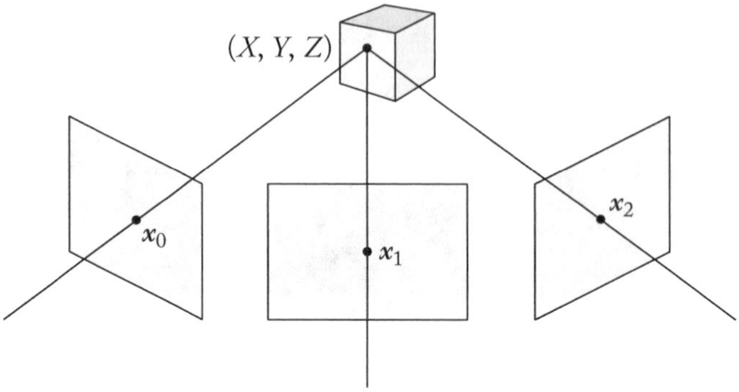

Figure 10.1: Trinocular stereo vision. A 3D point (X, Y, Z) is defined if the rays of points $\boldsymbol{x}_0$, $\boldsymbol{x}_1$, and $\boldsymbol{x}_2$ intersect at one point in the scene.

For the standard perspective projection cameras, the 3D point (X, Y, Z) and the three points $\boldsymbol{x}_0$, $\boldsymbol{x}_1$, and $\boldsymbol{x}_2$ are related by

$$\begin{pmatrix} x_\kappa/f_0 \\ y_\kappa/f_0 \\ 1 \end{pmatrix} \simeq \boldsymbol{P}_\kappa \begin{pmatrix} X \\ Y \\ Z \\ 1 \end{pmatrix}, \qquad \kappa = 0, 1, 2, \tag{10.2}$$

where $\boldsymbol{P}_\kappa$ are some 3×4 matrices, called the *camera matrices*, and the symbol $\simeq$ denotes equality up to an unknown scale ($\hookrightarrow$ Eq. (9.16)). The camera matrix $\boldsymbol{P}_\kappa$ encodes the position, orientation, and internal parameters, including the focal length, of the κth camera [5, 12].

10.2 TRIFOCAL TENSOR

The three points $\boldsymbol{x}_\kappa$, $\kappa = 0, 1, 2$, of Eq. (10.1) define a point in the scene if and only if their rays, or lines of sight, intersect at one in the scene (Fig. 10.1). We now derive that condition.

Eliminating the proportionality constant implied by $\simeq$ in Eq, (10.2), we obtain the fractional equations

$$\begin{aligned} x_\kappa &= f_0 \frac{P_{\kappa(11)}X + P_{\kappa(12)}Y + P_{\kappa(13)}Z + P_{\kappa(14)}}{P_{\kappa(31)}X + P_{\kappa(32)}Y + P_{\kappa(33)}Z + P_{\kappa(34)}}, \\ y_\kappa &= f_0 \frac{P_{\kappa(21)}X + P_{\kappa(22)}Y + P_{\kappa(23)}Z + P_{\kappa(24)}}{P_{\kappa(31)}X + P_{\kappa(32)}Y + P_{\kappa(33)}Z + P_{\kappa(34)}}, \end{aligned} \tag{10.3}$$

where $P_{\kappa(ij)}$ denotes the (i, j) element of $\boldsymbol{P}_\kappa$. Clearing fractions of the first equation, we obtain

$$\begin{aligned} & x_\kappa P_{\kappa(31)}X + x_\kappa P_{\kappa(32)}Y + x_\kappa P_{\kappa(33)}Z + x_\kappa P_{\kappa(34)} \\ &= f_0 P_{\kappa(11)}X + f_0 P_{\kappa(12)}Y + f_0 P_{\kappa(13)}Z + f_0 P_{\kappa(14)}. \end{aligned} \tag{10.4}$$

We do the same for the second equation of Eq. (10.3) and rearrange the resulting equations for $\kappa = 0, 1, 2$. In the end, we obtain the following linear equations:

$$\begin{pmatrix} f_0 P_{0(11)} - x_0 P_{0(31)} & f_0 P_{0(12)} - x_0 P_{0(32)} & f_0 P_{0(13)} - x_0 P_{0(33)} & f_0 P_{0(14)} - x_0 P_{0(34)} \\ f_0 P_{0(21)} - y_0 P_{0(31)} & f_0 P_{0(22)} - y_0 P_{0(32)} & f_0 P_{0(23)} - y_0 P_{0(33)} & f_0 P_{0(24)} - y_0 P_{0(34)} \\ f_0 P_{1(11)} - x_1 P_{1(31)} & f_0 P_{1(12)} - x_1 P_{1(32)} & f_0 P_{1(13)} - x_1 P_{1(33)} & f_0 P_{1(14)} - x_1 P_{1(34)} \\ f_0 P_{1(21)} - y_1 P_{1(31)} & f_0 P_{1(22)} - y_1 P_{1(32)} & f_0 P_{1(23)} - y_1 P_{1(33)} & f_0 P_{1(24)} - y_1 P_{1(34)} \\ f_0 P_{2(11)} - x_2 P_{2(31)} & f_0 P_{2(12)} - x_2 P_{2(32)} & f_0 P_{2(13)} - x_2 P_{2(33)} & f_0 P_{2(14)} - x_2 P_{2(34)} \\ f_0 P_{2(21)} - y_2 P_{2(31)} & f_0 P_{2(22)} - y_2 P_{2(32)} & f_0 P_{2(23)} - y_2 P_{2(33)} & f_0 P_{2(24)} - y_2 P_{2(34)} \end{pmatrix} \begin{pmatrix} X \\ Y \\ Z \\ 1 \end{pmatrix} = \begin{pmatrix} 0 \\ 0 \\ 0 \\ 0 \\ 0 \\ 0 \end{pmatrix}. \tag{10.5}$$

This equation has a unique solution (X, Y, Z) if and only if three among its six equations are linearly independent and others are written as their linear combinations, or equivalently the matrix on the left side has rank 3. This means that its arbitrary 4×4 minors are 0. We can obtain this constraint by extracting, from among the six rows, four rows that contain quantities of the three images and letting the resulting minor be 0. In the end, we obtain

$$\sum_{i,j,k,l,m=1}^{3} \epsilon_{ljp}\epsilon_{mkq} T_i^{lm} x_{0(i)} x_{1(j)} x_{2(k)} = 0, \qquad p, q = 1, 2, 3, \tag{10.6}$$

where ϵ_{ijk} is the permutation signature, taking 1 if (i, j, k) is an even permutation of (1,2,3), -1 if it is an odd permutation, and 0 otherwise ($\hookrightarrow$ Problem 10.1). Here, $x_{\kappa(i)}$ denotes the ith component of vector $\boldsymbol{x}_\kappa$. Equation (10.6) is called the *trilinear constraint*, and the coefficients T_i^{lm} are called the *trifocal tensor*; they are determined by the relative configuration of the three cameras. Given three camera matrices $\boldsymbol{P}_0$, $\boldsymbol{P}_1$, and $\boldsymbol{P}_2$, the trifocal tensor has the form

$$
T_1^{jk} = \begin{vmatrix} P_{0(21)} & P_{0(22)} & P_{0(23)} & P_{0(24)} \\ P_{0(31)} & P_{0(32)} & P_{0(33)} & P_{0(34)} \\ P_{1(j1)} & P_{1(j2)} & P_{1(j3)} & P_{1(j4)} \\ P_{2(k1)} & P_{2(k2)} & P_{2(k3)} & P_{2(k4)} \end{vmatrix},
$$

$$
T_2^{jk} = \begin{vmatrix} P_{0(31)} & P_{0(32)} & P_{0(33)} & P_{0(34)} \\ P_{0(11)} & P_{0(12)} & P_{0(13)} & P_{0(14)} \\ P_{1(j1)} & P_{1(j2)} & P_{1(j3)} & P_{1(j4)} \\ P_{2(k1)} & P_{2(k2)} & P_{2(k3)} & P_{2(k4)} \end{vmatrix}, \tag{10.7}
$$

$$
T_3^{jk} = \begin{vmatrix} P_{0(11)} & P_{0(12)} & P_{0(13)} & P_{0(14)} \\ P_{0(21)} & P_{0(22)} & P_{0(23)} & P_{0(24)} \\ P_{1(j1)} & P_{1(j2)} & P_{1(j3)} & P_{1(j4)} \\ P_{2(k1)} & P_{2(k2)} & P_{2(k3)} & P_{2(k4)} \end{vmatrix},
$$

where $P_{\kappa(ij)}$ denotes the (i, j) element of the matrix $\boldsymbol{P}_\kappa$. Using the permutation signature, these three expressions are written as the single form

$$
T_i^{jk} = \sum_{l,m=1}^{3} \epsilon_{ilm} |\boldsymbol{P}_{0012}^{lmjk}|, \tag{10.8}
$$

where $\boldsymbol{P}_{0012}^{lmjk}$ is a 4×4 matrix consisting of the lth row of $\boldsymbol{P}_0$, the mth row of $\boldsymbol{P}_0$, the jth row of $\boldsymbol{P}_1$, and the kth row of $\boldsymbol{P}_2$ as its rows in that order. The nine equations of Eq. (10.6) for $p, q = 1, 2, 3$ are not linearly independent; only four are independent (↪ Problem 10.2).

10.3 OPTIMAL CORRECTION OF CORRESPONDENCES

If we observe a point P in the scene at $\boldsymbol{x}_0$, $\boldsymbol{x}_1$, and $\boldsymbol{x}_2$ in the three camera images as in Fig. 10.1, they should ideally satisfy the trifocal constraint of Eq. (10.6), but that does not necessarily hold for real images. This is because the three points $\boldsymbol{x}_0$, $\boldsymbol{x}_1$, and $\boldsymbol{x}_2$ are obtained by image processing; we search different images for those pixels whose neighborhoods are similar to each other. Many algorithms have been proposed for extracting matching points over different images, but any algorithm entails some degree of uncertainty.

Here, we consider the problem of correcting the three points $\boldsymbol{x}_0$, $\boldsymbol{x}_1$, and $\boldsymbol{x}_2$ to $\hat{\boldsymbol{x}}_0$, $\hat{\boldsymbol{x}}_1$, and $\hat{\boldsymbol{x}}_2$ so that the trifocal constraint of Eq. (10.6) is exactly satisfied. We assume that the matching algorithm is fairly accurate so that the uncertainty is very small and write

$$
\boldsymbol{x}_\kappa = \bar{\boldsymbol{x}}_\kappa + \Delta \boldsymbol{x}_\kappa, \qquad \kappa = 0, 1, 2, \tag{10.9}
$$

where $\bar{\boldsymbol{x}}_\kappa$ are the true positions that satisfy Eq. (10.6). We correct $\boldsymbol{x}_0$, $\boldsymbol{x}_1$, and $\boldsymbol{x}_2$ in an optimal way in the sense of least squares. Namely, we estimate the true positions $\bar{\boldsymbol{x}}_\kappa$ that minimize the sum of the squares, called the *reprojection error*,

$$E = \sum_{\kappa=0}^{2} \|\boldsymbol{x}_\kappa - \bar{\boldsymbol{x}}_\kappa\|^2, \tag{10.10}$$

subject to the constraint that $\bar{\boldsymbol{x}}_\kappa$ satisfy Eq. (10.6).

We substitute $\bar{\boldsymbol{x}}_\kappa = \boldsymbol{x}_\kappa - \Delta\boldsymbol{x}_\kappa$ into Eq. (10.10), ignore high-order terms in $\Delta\boldsymbol{x}_\kappa$, and compute the value of $\Delta\boldsymbol{x}_\kappa$ that minimizes E. Write the resulting value $\Delta\hat{\boldsymbol{x}}_\kappa$. Substituting $\bar{\boldsymbol{x}}_\kappa = \boldsymbol{x}_\kappa - \Delta\hat{\boldsymbol{x}}_\kappa$ into Eq. (10.10) and ignoring high-order terms in $\Delta\hat{\boldsymbol{x}}_\kappa$, we compute the value of $\Delta\hat{\boldsymbol{x}}_\kappa$ that minimizes E. Writing the resulting value as $\Delta\hat{\hat{\boldsymbol{x}}}_\kappa$, we substitute $\bar{\boldsymbol{x}}_\kappa = \boldsymbol{x}_\kappa - \Delta\hat{\hat{\boldsymbol{x}}}_\kappa$ into Eq. (10.10) and repeat this process until the reproduction error E no longer changes. The procedure is summarized as follows (↪ Problem 10.3).

1. Let $E_0 = \infty$ (a sufficiently large number), $\hat{\boldsymbol{x}}_\kappa = \boldsymbol{x}_\kappa$, and $\tilde{\boldsymbol{x}}_\kappa = \boldsymbol{0}$, $\kappa = 0, 1, 2$.

2. Compute the following P_{pqs}, Q_{pqs}, and R_{pqs}, $p, q, s = 1, 2, 3$:

$$\begin{aligned} P_{pqs} &= \sum_{i,j,k,l,m=1}^{3} \epsilon_{ljp}\epsilon_{mkq} T_i^{lm} P_{\boldsymbol{k}(si)} \hat{x}_{1(j)} \hat{x}_{2(k)}, \\ Q_{pqs} &= \sum_{i,j,k,l,m=1}^{3} \epsilon_{ljp}\epsilon_{mkq} T_i^{lm} \hat{x}_{0(i)} P_{\boldsymbol{k}(sj)} \hat{x}_{2(k)}, \\ R_{pqs} &= \sum_{i,j,k,l,m=1}^{3} \epsilon_{ljp}\epsilon_{mkq} T_i^{lm} \hat{x}_{0(i)} \hat{x}_{1(j)} P_{\boldsymbol{k}(sk)}, \end{aligned} \tag{10.11}$$

   where $P_{\boldsymbol{k}(ij)}$ is the (i, j) element of the projection matrix $\boldsymbol{P_k} = \boldsymbol{I} - \boldsymbol{k}\boldsymbol{k}^\top$ along $\boldsymbol{k} = (0, 0, 1)^\top$ (↪ Eq. (2.17)).

3. Compute the following C_{pqrs} and F_{pq}:

$$\begin{aligned} C_{pqrs} &= \sum_{i,j,k,l,m=1}^{3} \epsilon_{ljp}\epsilon_{mkq} T_i^{lm} \Big(P_{rsi} \hat{x}_{1(j)} \hat{x}_{2(k)} + \hat{x}_{0(i)} Q_{rsj} \hat{x}_{2(k)} + \hat{x}_{0(i)} \hat{x}_{1(j)} R_{rsk} \Big), \\ F_{pq} &= \sum_{i,j,k,l,m=1}^{3} \epsilon_{ljp}\epsilon_{mkq} T_i^{lm} \Big(\hat{x}_{0(i)} \hat{x}_{1(j)} \hat{x}_{2(k)} + \tilde{x}_{0(i)} \hat{x}_{1(j)} \hat{x}_{2(k)} + \hat{x}_{0(i)} \tilde{x}_{1(j)} \hat{x}_{2(k)} \\ &\quad + \hat{x}_{0(i)} \hat{x}_{1(j)} \tilde{x}_{2(k)} \Big). \end{aligned} \tag{10.12}$$

4. Solve the following linear equation, using pseudoinverse of truncated rank 3, to compute λ_{pq}:

$$\sum_{r,s=1}^{3} C_{pqrs} \lambda_{rs} = F_{pq}, \qquad p, q = 1, 2, 3. \tag{10.13}$$

5. Update $\tilde{\boldsymbol{x}}_\kappa$ and $\hat{\boldsymbol{x}}_\kappa$, $\kappa = 0, 1, 2$, to

$$\tilde{x}_{0(i)} \leftarrow \sum_{p,q=1}^{3} P_{pqi}\lambda_{pq}, \quad \tilde{x}_{1(i)} \leftarrow \sum_{p,q=1}^{3} Q_{pqi}\lambda_{pq}, \quad \tilde{x}_{2(i)} \leftarrow \sum_{p,q=1}^{3} R_{pqi}\lambda_{pq},$$

$$\hat{\boldsymbol{x}}_\kappa \leftarrow \boldsymbol{x}_\kappa - \tilde{\boldsymbol{x}}_\kappa. \tag{10.14}$$

6. Compute the following reprojection error E:

$$E = \sum_{\kappa=0}^{2} \|\tilde{\boldsymbol{x}}_\kappa\|^2. \tag{10.15}$$

7. If $E \approx E_0$, return E and $\hat{\boldsymbol{x}}_\kappa$, $\kappa = 0, 1, 2$, and stop. Otherwise, let $E_0 \leftarrow E$ and go back to Step 2.

10.4 SOLVING LINEAR EQUATIONS

By "solve Eq. (10.13) using pseudoinverse of truncated rank 3" in Step 4 of the above procedure, we mean the following. The unknowns λ_{pq} of Eq. (10.13) are the Lagrange multipliers for minimizing Eq. (10.10). In matrix form, we can write (10.13) as

$$\begin{pmatrix} C_{1111} & C_{1112} & C_{1113} & C_{1121} & C_{1122} & C_{1123} & C_{1131} & C_{1132} & C_{1133} \\ C_{1211} & C_{1212} & C_{1213} & C_{1221} & C_{1222} & C_{1223} & C_{1231} & C_{1232} & C_{1233} \\ C_{1311} & C_{1312} & C_{1313} & C_{1321} & C_{1322} & C_{1323} & C_{1331} & C_{1332} & C_{1333} \\ C_{2111} & C_{2112} & C_{2113} & C_{2121} & C_{2122} & C_{2123} & C_{2131} & C_{2132} & C_{2133} \\ C_{2211} & C_{2212} & C_{2213} & C_{2221} & C_{2222} & C_{2223} & C_{2231} & C_{2232} & C_{2233} \\ C_{2311} & C_{2312} & C_{2313} & C_{2321} & C_{2322} & C_{2323} & C_{2331} & C_{2332} & C_{2333} \\ C_{3111} & C_{3112} & C_{3113} & C_{3121} & C_{3122} & C_{3123} & C_{3131} & C_{3132} & C_{3133} \\ C_{3211} & C_{3212} & C_{3213} & C_{3221} & C_{3222} & C_{3223} & C_{3231} & C_{3232} & C_{3233} \\ C_{3311} & C_{3312} & C_{3313} & C_{3321} & C_{3322} & C_{3323} & C_{3331} & C_{3332} & C_{3333} \end{pmatrix} \begin{pmatrix} \lambda_{11} \\ \lambda_{12} \\ \lambda_{13} \\ \lambda_{21} \\ \lambda_{22} \\ \lambda_{23} \\ \lambda_{31} \\ \lambda_{32} \\ \lambda_{33} \end{pmatrix}$$

$$= \begin{pmatrix} F_{11} \\ F_{12} \\ F_{13} \\ F_{21} \\ F_{22} \\ F_{23} \\ F_{31} \\ F_{32} \\ F_{33} \end{pmatrix}. \tag{10.16}$$

However, *the coefficient matrix has rank 6*, only six columns and six rows being independent. This rank deficiency is due to fact that we are correcting six components Δx_0, Δy_0, Δx_1, Δy_1, Δx_2, and Δx_2, while there exist nine equations. This is because we regard all the nine components of the vectors $\Delta\boldsymbol{x}_0$, $\Delta\boldsymbol{x}_1$, and $\Delta\boldsymbol{x}_2$ as unknowns, while the third components of $\boldsymbol{x}_\kappa$ are all constant 1 and hence the third components of $\Delta\boldsymbol{x}_\kappa$ are all 0. In principle, we could obtain a unique solution if we select six from the nine equation of Eq. (10.16).

However, there exists an additional problem. If vectors $\boldsymbol{x}_\kappa$ exactly satisfy Eq. (10.6), *the rank decreases to 3*. This is because while Eq. (10.6) consists of nine equations for p, $q = 1$, 2, 3, *only three of them are independent*. This is understood as follows. Each equation of Eq. (10.6) defines a cubic polynomial hypersurface in the 6D space of x_0, y_0, x_1, y_1, x_2, and x_2, and the solution of Eq. (10.6) is the intersection of the resulting nine hypersurfaces. *The intersection should have dimension 3*, because the solution must correspond to positions where the three rays meet at a point. In other words, points in 3D and triplets of their image points over three views must correspond one to one. In 6D, the intersection of three hypersurfaces has generally dimension 3. Hence, if nine hypersurfaces intersect, only three of them are independent; the rest are redundant. Of course, the observed values and intermediate values in the correction computation do not exactly satisfy Eq. (10.6), so the matrix in Eq. (10.16) generally has rank 6. If we select six equations from Eq. (10.16), they degenerate to rank 3 in the limit of convergence, so the numerical computation becomes unstable (i.e., rounding errors are magnified) as it approaches convergence. This can be handled by selecting from Eq. (10.16) the three equations that are the "most independent." This is equivalent to computing the singular value decomposition and solving those three equations associated with the largest three singular values, i.e., using the pseudoinverse of truncated rank 3. Specifically, we write Eq. (10.16) as

$$\boldsymbol{C}\boldsymbol{\lambda} = \boldsymbol{f}, \tag{10.17}$$

and compute $\boldsymbol{\lambda}$ by

$$\boldsymbol{\lambda} = \boldsymbol{C}_3^{-}\boldsymbol{f}. \tag{10.18}$$

As pointed out in Chapter 6, this is also equivalent to solving Eq. (10.16) by least squares.

10.5 EFFICIENCY OF COMPUTATION

The above computation requires repeated evaluation of expressions of the form $\boldsymbol{T}(\boldsymbol{x}, \boldsymbol{y}, \boldsymbol{z})$, which for three input vectors $\boldsymbol{x} = (x_i)$, $\boldsymbol{y} = (y_i)$, and $\boldsymbol{z} = (z_i)$ outputs the following matrix $\boldsymbol{T} = (T_{pq})$:

$$T_{pq} = \sum_{i,j,k,l,m=1}^{3} \epsilon_{ljp}\epsilon_{mkq} T_i^{lm} x_i y_j z_k. \tag{10.19}$$

The right side is the sum of $3^5 = 243$ terms, which is computed 9 times for p, $q = 1$, 2, 3, requiring 2,187 summations in total. This is a time consuming process but can be made efficient by rewriting Eq. (10.19) in the form

$$T_{pq} = \frac{1}{4}\sum_{i,j,k,l,m=1}^{3} \epsilon_{ljp}\epsilon_{mkq}x_i\Big(T_i^{lm}y_jz_k - T_i^{jm}y_lz_k - T_i^{lk}y_jz_m + T_i^{jk}y_lz_m\Big). \tag{10.20}$$

From the properties of the permutation signature ϵ_{ijk}, we can easily confirm that each term in the sum and of $\sum_{i,j,k,l=1}^{3}$ equals Eq. (10.19). We can also see that the expression is symmetric with respect to l and j, e.g., the term for $l = 1$ and $j = 2$ is equal to the term for $l = 2$ and $j = 1$. Hence, we only need to compute either of them and multiply it by 2. Since all terms are multiplied by ϵ_{ljp} and summed, we need not consider those terms for which l or j equals p. In other words, we only need to sum over l and j not equal to p such that $\epsilon_{ljp} = 1$. Similarly, we only need to sum over m and k not equal to q such that $\epsilon_{mkq} = 1$. This is systematically done by "addition $\oplus$ modulo 3": $1 \oplus 1 = 2$, $1 \oplus 2 = 3$, $3 \oplus 1 = 1$, etc. Since $\epsilon_{p\oplus1,p\oplus2,p} = 1$ and $\epsilon_{q\oplus1,q\oplus2,q} = 1$, Eq. (10.20) is equivalently written as

$$\begin{aligned} T_{pq} \;&=\; \sum_{i=1}^{3} x_i\Big(T_i^{p\oplus1,q\oplus1}y_{p\oplus2}z_{q\oplus2} - T_i^{p\oplus2,q\oplus1}y_{p\oplus1}z_{q\oplus2} - T_i^{p\oplus1,q\oplus2}y_{p\oplus2}z_{q\oplus1} \\ &\quad + T_i^{p\oplus2,q\oplus2}y_{p\oplus1}z_{q\oplus1}\Big). \end{aligned} \tag{10.21}$$

In this form, only 12 terms are summed, reducing the number of additions to about 1/20 as compared with Eq. (10.19). The operation $\oplus$ can be defined as an inline function in the program.

10.6 3D POSITION COMPUTATION

Once the observed three points have been corrected so that they satisfy the trilinear constraint of Eq. (10.6), which is the necessary and sufficient condition that their rays meet at a common intersection in the scene, the 3D position (X, Y, Z) is computed by solving Eq. (10.5), which is rewritten as

$$\boldsymbol{T}\begin{pmatrix} X \\ Y \\ Z \end{pmatrix} = -\boldsymbol{p}, \tag{10.22}$$

where

$$\boldsymbol{T} = \begin{pmatrix} f_0P_{0(11)} - x_0P_{0(31)} & f_0P_{0(12)} - x_0P_{0(32)} & f_0P_{0(13)} - x_0P_{0(33)} \\ f_0P_{0(21)} - y_0P_{0(31)} & f_0P_{0(22)} - y_0P_{0(32)} & f_0P_{0(23)} - y_0P_{0(33)} \\ f_0P_{1(11)} - x_1P_{1(31)} & f_0P_{1(12)} - x_1P_{1(32)} & f_0P_{1(13)} - x_1P_{1(33)} \\ f_0P_{1(21)} - y_1P_{1(31)} & f_0P_{1(22)} - y_1P_{1(32)} & f_0P_{1(23)} - y_1P_{1(33)} \\ f_0P_{2(11)} - x_2P_{2(31)} & f_0P_{2(12)} - x_2P_{2(32)} & f_0P_{2(13)} - x_2P_{2(33)} \\ f_0P_{2(21)} - y_2P_{2(31)} & f_0P_{2(22)} - y_2P_{2(32)} & f_0P_{2(23)} - y_2P_{2(33)} \end{pmatrix}, \tag{10.23}$$

$$p = \begin{pmatrix} f_0 P_{0(14)} - x_0 P_{0(34)} \\ f_0 P_{0(24)} - y_0 P_{0(34)} \\ f_0 P_{1(14)} - x_1 P_{1(34)} \\ f_0 P_{1(24)} - y_1 P_{1(34)} \\ f_0 P_{2(14)} - x_2 P_{2(34)} \\ f_0 P_{2(24)} - y_2 P_{2(34)} \end{pmatrix}. \tag{10.24}$$

From among the six component equations, we can pick out any three, and the solution does not depend on the choice of which three, since a unique solution is guaranteed by the trilinear constraint.

In practice, however, it is much simpler to solve Eq. (10.22) by least squares using all the six equations; the solution is the same. As pointed out in Chapter 6, this is equivalent to using the pseudoinverse:

$$\begin{pmatrix} X \\ Y \\ Z \end{pmatrix} = -\boldsymbol{T}^{-} \boldsymbol{p}. \tag{10.25}$$

This has the added advantage that we can obtain an approximation of X, Y, and Z, even if the three rays do not meet at a point.

GLOSSARY AND SUMMARY

camera matrix: For the standard "perspective projection" cameras, the homogeneous coordinates of a point in the camera image is obtained from the homogeneous coordinates of the 3D point we are viewing by multiplying a 3×4 "camera matrix" $\boldsymbol{P}$ (Eq. (10.2)). Note the scale indeterminacy of homogeneous coordinates.

homogeneous coordinates: Representation of 2D point by three numbers or a 3D point by four numbers. It is obtained by padding the usual ("inhomogeneous") coordinates with 1, but only the ratios among homogeneous coordinates have geometric meaning. If the last coordinate is 0, the corresponding point is interpreted to be infinitely far away.

optimal correction: Correcting three points on three camera images so that they correspond to an identical point in the scene, i.e., the trilinear constraint is satisfied. The correction is done in such a way that the sum of the square lengths of the displacement, called the "reprojection error" (Eq. (10.10)), is minimized.

triangulation: The process of computing the 3D positions using multiple camera images. The camera setting for this is called "stereo vision." It is "binocular" and "trinocular," respectively, if two and three cameras are used.

trifocal tensor: The set of coefficients that appear in the trilinear constraint. They have three indices and determined by the positions, orientations, and internal parameters of the three cameras that view the scene.

trilinear constraint: The necessary and sufficient condition for the rays (or the "lines of sight") associated with three points on three camera images intersect at a point in the scene (Eq. (10.6)). It consists of nine equations for p, $q = 1, 2, 3$, but only four are linearly independent.

- Points (x_0, y_0), (x_1, y_1), and (x_2, y_2) in three camera images correspond to an identical point in the scene if and only the coefficient matrix in Eq. (10.5) has rank 3, meaning that only three among the six equations are linearly independent. This leads to the trilinear constraint of Eq. (10.6) and the expression of the trifocal tensor of Eq. (10.7).

- Each of the nine equations of the trilinear constraint of Eq. (10.6) defines a 5D hypersurface in the 6D space of (x_0, y_0), (x_1, y_1), and (x_2, y_2). However, their intersection must be a 3D space, because it corresponds to the viewed point (X, Y, Z) one to one. This means that the intersection is defined only by three of the nine hypersurfaces; the rest automatically pass through it.

- Optimal correction of three-view correspondence is done by iteratively computing the displacements $\Delta \boldsymbol{x}_\kappa$ of $\boldsymbol{x}_\kappa$, $\kappa = 0, 1, 2$, so that the reprojection error E of Eq. (10.10) decreases. In each stage, linear equations are solved by ignoring high-order terms of $\Delta \boldsymbol{x}_\kappa$, and this process is iterated.

- In iteratively solving linear equations for optimal correction, we need to take into consideration the degeneracy structure of the trilinear constraint of Eq. (10.6). This is taken care of by computing rank-constrained pseudoinverse: we constrain the coefficient matrix of Eq. (10.16) to rank 3.

- In evaluating terms involving the trifocal tensor, 243 operations are required for summation over four indices, but the number of operations reduces to 12 by considering the symmetry of the indices. This is most succinctly done by modulo 3 computation (Eq. (10.21)).

- If the three corresponding image points satisfy the trilinear constraint, they define a unique 3D point. In practice, the use of pseudoinverse (Eq. (10.25)) is the most convenient for this computation.

10.7 SUPPLEMENTAL NOTES

In this chapter, we considered triangulation from three views, but historically triangulation, i.e., determination of 3D positions from camera images, has been based on two views in the name of *stereo vision*. It is also called *binocular stereo vision* when the use of two cameras is emphasized.

Let $\boldsymbol{x}_1$ and $\boldsymbol{x}_2$ be the corresponding points represented as 3D vectors as in Eq. (10.1). It is easily seen [5, 12] that the condition for their rays to meet at a point in the scene is given by

$$\langle \boldsymbol{x}_1, \boldsymbol{F}\boldsymbol{x}_2 \rangle = 0, \tag{10.26}$$

where $\boldsymbol{F}$ is a 3×3 matrix, called the *fundamental matrix*, which encodes the positions, orientations, and internal parameters of the two cameras. Equation (10.26) is called the *epipolar constraint*. In terms of the perspective projection modeling of Eq. (10.2), the elements of the fundamental matrix $\boldsymbol{F}$ is expressed in terms of the camera matrices $\boldsymbol{P}_1$ and $\boldsymbol{P}_2$ of the two cameras in the form

$$F_{11} = \begin{vmatrix} P_{1(21)} & P_{1(22)} & P_{1(23)} & P_{1(24)} \\ P_{1(31)} & P_{1(32)} & P_{1(33)} & P_{1(34)} \\ P_{2(21)} & P_{2(22)} & P_{2(23)} & P_{2(24)} \\ P_{2(31)} & P_{2(32)} & P_{2(33)} & P_{2(34)} \end{vmatrix}, \quad \ldots, \quad F_{33} = \begin{vmatrix} P_{1(11)} & P_{1(12)} & P_{1(13)} & P_{1(14)} \\ P_{1(21)} & P_{1(22)} & P_{1(23)} & P_{1(24)} \\ P_{2(11)} & P_{2(12)} & P_{2(13)} & P_{2(14)} \\ P_{2(21)} & P_{2(22)} & P_{2(23)} & P_{2(24)} \end{vmatrix}. \tag{10.27}$$

Using the permutation symbol, these nine elements are expressed as a single form

$$F_{ij} = \sum_{k,l,m,n=1}^{3} \epsilon_{ijk}\epsilon_{jmn} |\boldsymbol{P}_{1122}^{klmn}|. \tag{10.28}$$

As pointed out in Section 10.3, the rays associated with the corresponding image points do not necessarily meet in the scene since image correspondence detection entails some degree of uncertainty. In the old days, this was handled by a practical compromise such as regarding the "midpoint" of the shortest line segment connecting the two rays as their intersection. However, Kanazawa and Kanatani [15] pointed out that observed points should be minimally corrected so that the epipolar equation of Eq. (10.26) is exactly satisfied. To do this, they adopted the statistical approach introduced in Chapter 7, viewing the detected points $\boldsymbol{x}_1$ and $\boldsymbol{x}_2$ as random variables. They minimized the *reprojection error*

$$E = \sum_{\kappa=1}^{2} \|\boldsymbol{x}_\kappa - \bar{\boldsymbol{x}}_\kappa\|^2, \tag{10.29}$$

for estimating $\bar{\boldsymbol{x}}_1$ and $\bar{\boldsymbol{x}}_2$ subject to the constraint that they satisfy the epipolar equation of Eq. (10.26). In statistical terms, this is *maximum likelihood estimation*, or *maximum likelihood* for short, on the assumption that the noise has independent, homogeneous, and isotropic normal (or Gaussian) distributions; noise is homogeneous if the distribution is the same for all points and isotropic if the distribution is symmetric in all directions. A probability density function is called the *likelihood* if the variables in it are replaced by their actual observations. If the noise in $\boldsymbol{x}_1$ and $\boldsymbol{x}_2$ has independent, homogeneous, and isotropic normal distributions, their likelihood is evidently a function of the reprojection error E of Eq. (10.29), so minimizing E is equivalent to maximizing the likelihood. Maximum likelihood is regarded as the standard method for statistical estimation.

If the noise in $\boldsymbol{x}_1$ and $\boldsymbol{x}_2$ is not isotropic (i.e., is anisotropic), having respective covariance matrices $V[\boldsymbol{x}_1]$ and $V[\boldsymbol{x}_2]$, maximum likelihood is equivalent to minimizing

$$E = \sum_{\kappa=1}^{2} \langle \boldsymbol{x}_\kappa - \bar{\boldsymbol{x}}_\kappa, V[\boldsymbol{x}_\kappa]^-(\boldsymbol{x}_\kappa - \bar{\boldsymbol{x}}_\kappa)\rangle. \tag{10.30}$$

Here, the pseudoinverse $V[\boldsymbol{x}_\kappa]^-$ appears, as discussed in Section 7.2, because the third components of $\boldsymbol{x}_1$ and $\boldsymbol{x}_2$ are fixed constant by definition ($\hookrightarrow$ Eq. (10.1)). The sum of square distances weighted by the inverse (or pseudoinverse) of covariance matrices is called *Mahalanobis distance*. Maximum likelihood for anisotropic noise generally reduces to Mahalanobis distance minimization. See [8] for more about statistical estimation of geometric problems. If the noise is homogeneous and isotropic, we can replace $V[\boldsymbol{x}_\kappa]$ by $\boldsymbol{P_k}$ ($\equiv \mathrm{diag}(1, 1, 0)$), and minimization of Eq. (10.30) is equivalent to minimization of Eq. (10.29).

What Kanazawa and Kanatani [15] showed was a first approximation, which is sufficient for most real applications, but later Kanatani et al. [13] gave an iterative procedure for computing the exact solution. Meanwhile, Hartley and Sturm [4] proposed a non-iterative direct algebraic method. Their method reduces the computation to six-degree polynomial solving. Both method compute the same solution. It has been shown, however, that the iterative method of Kanatani et al. [13] is much faster; the direct method of [4] spends much time on six-degree polynomial solving.

The iterative method for three-view triangulation described in this chapter is based on Kanatani et al. [14]. It can also be viewed as maximum likelihood estimation. Namely, if the noise terms $\Delta\boldsymbol{x}_\kappa$ in Eq. (10.9) are regarded as random variables subject to an independent, homogeneous, and isotropic normal distribution, minimizing Eq. (10.10) is equivalent to maximizing the likelihood. There also exists a direct algebraic method for three-view triangulation, solving a 47-degree polynomial [20], but it does not have a practical meaning due to computational burden.

The formulation in this chapter can be extended to four-view triangulation. Let $\boldsymbol{x}_\kappa$, $\kappa = 0$, 1, 2, 3, be points in four camera images viewing a point in the scene. The necessary and sufficient condition for the associated rays intersect at a point in the scene is given, instead of Eq. (10.6), by

$$\sum_{i,j,k,l,m,n,p,q=1}^{3} \epsilon_{ima}\epsilon_{jnb}\epsilon_{kpc} Q^{ijkl} x_{0(m)} x_{1(n)} x_{2(p)} x_{3(q)} = 0, \quad a, b, c = 1, 2, 3, \tag{10.31}$$

where Q^{ijkl} is called the *quadrifocal tensor*, whose components are given, instead of Eq. (10.8), by

$$Q^{ijkl} = |\boldsymbol{P}_{0123}^{ijkl}|. \tag{10.32}$$

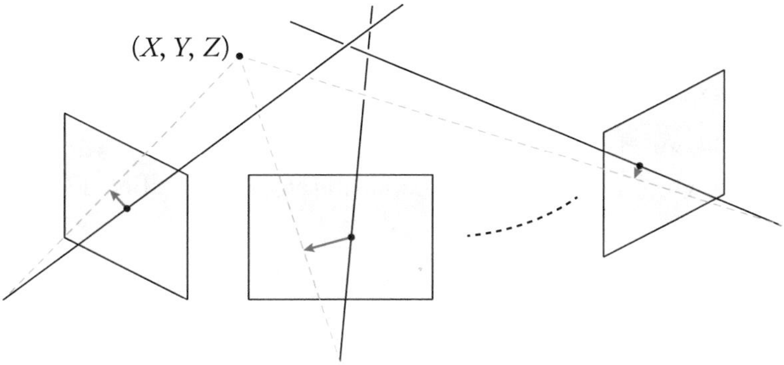

Figure 10.2: Correcting observed points so that their rays meet in the scene in such a way that the reprojection error (= the sum of square distances of displacements) is minimized.

Equation (10.31) is called the *quadrilinear constraint*. As in the two- and three-view cases, a direct algebraic method is known to exist.[1] However, Kanatani et al. [14] showed that triangulation from four and more views can be solved by straightforwardly extending the three-view triangulation to M views (Fig. 10.2). Namely, we adopt the statistical approach to correct $\boldsymbol{x}_\kappa$, $\kappa = 0, \ldots, M-1$, by maximum likelihood, minimizing the reprojection error,

$$E = \sum_{\kappa=0}^{M-1} \|\boldsymbol{x}_\kappa - \bar{\boldsymbol{x}}_\kappa\|^2, \tag{10.33}$$

subject to the condition that the trilinear constraint is satisfied by "consecutive three image triplets," i.e.,

$$\sum_{i,j,k,l,m=1}^{3} \epsilon_{ljp}\epsilon_{mkq} T_{(\kappa)i}^{lm} x_{\kappa(i)} x_{\kappa+1(j)} x_{\kappa+2(k)} = 0, \qquad \kappa = 0, \ldots, M-3, \tag{10.34}$$

The optimization goes just as in the three-view case; again, pseudoinverse plays a crucial role for dealing with the degenerate structure. See [12] for concrete computational procedures and real image examples.

10.8 PROBLEMS

10.1. Show that the trilinear constraint of Eq. (10.6) and the trifocal tensor T_i^{jk} of Eq. (10.7) are obtained from the condition that Eq. (10.5) has a unique solution.

[1]It is known that optimal triangulation is algebraically computed from four-, five-, six-, and seven-views, at least theoretically, by solving a 148-. a 336-, a 638-, and a 1081-degree polynomial, respectively [3].

10.2. The trilinear constraint of Eq. (10.6) consists of nine equations for $p, q = 1, 2, 3$. Show that only four of them are linearly independent, i.e., the remaining ones are expressed as their linear combinations.

10.3. Show that the optimal correction of three-view correspondence is done by the procedure through Eqs. (10.11)–(10.15) as described at the end of Section 10.3.

APPENDIX A

Fundamentals of Linear Algebra

Here, we summarize fundamentals of linear algebra and related mathematical facts relevant to the discussions in the preceding chapters. First, we give a formal proof that a linear mapping is written as the product with some matrix. Then, we describe basic facts about the inner product and the norm of vectors. Next, we list formulas that frequently appear in relation to "linear forms," "quadratic forms," and "bilinear forms." Then, we briefly discuss expansion of vectors with respect to an orthonormal basis and least-squares approximation. We also introduce "Lagrange's method of indeterminate multipliers" for computing the maximum/minimum of a function of vectors subject to constraints. Finally, we summarize basics of eigenvalues and eigenvectors and their applications to maximization/minimization of a quadratic form.

A.1 LINEAR MAPPINGS AND MATRICES

A mapping $f(\cdot)$ from the nD space $\mathcal{R}^n$ to the mD space $\mathcal{R}^m$ is a *linear mapping* if

$$f(\boldsymbol{u}+\boldsymbol{v}) = f(\boldsymbol{u}) + f(\boldsymbol{v}), \qquad f(c\boldsymbol{u}) = cf(\boldsymbol{u}), \tag{A.1}$$

for arbitrary $\boldsymbol{u}, \boldsymbol{v} \in \mathcal{R}^n$ and an arbitrary real number c, i.e., if *a sum corresponds to a sum and a constant multiple corresponds to a constant multiple.*

Suppose a linear mapping $f(\cdot)$ maps a vector $\boldsymbol{u} \in \mathcal{R}^n$ to a vector $\boldsymbol{u}' \in \mathcal{R}^m$. If $\boldsymbol{u}$ is a column vector whose ith component is u_i, which we abbreviate to $\boldsymbol{u} = \left(u_i\right)$, we can write it as

$$\boldsymbol{u} = \sum_{j=1}^{n} u_j \boldsymbol{e}_j, \tag{A.2}$$

where $\boldsymbol{e}_j$ is the nD vector whose jth component is 1 and whose other components are all 0. We call the set $\{\boldsymbol{e}_1, \ldots, \boldsymbol{e}_n\}$ the *natural* (or *standard basis* of $\mathcal{R}^n$. Similarly, we can write the vector $\boldsymbol{u}' = \left(u'_i\right) \in \mathcal{R}^m$ as

$$\boldsymbol{u}' = \sum_{i=1}^{m} u'_i \boldsymbol{e}'_i, \tag{A.3}$$

using the natural basis $\{\boldsymbol{e}'_1, \ldots, \boldsymbol{e}'_m\}$ of $\mathcal{R}^m$.

The mapping of $\boldsymbol{u} \in \mathcal{R}^n$ to $\boldsymbol{u}' \in \mathcal{R}^m$ by $f(\cdot)$ is written as

$$\boldsymbol{u}' = f(\boldsymbol{u}) = f(\sum_{j=1}^{n} u_j \boldsymbol{e}_j) = \sum_{j=1}^{n} u_j f(\boldsymbol{e}_j). \tag{A.4}$$

Since $f(\boldsymbol{e}_j)$ is a vector of $\mathcal{R}^m$, it is expressed as a linear combination of the natural basis $\{\boldsymbol{e}'_1, \ldots, \boldsymbol{e}'_m\}$ of $\mathcal{R}^m$ in the form

$$f(\boldsymbol{e}_j) = \sum_{i=1}^{m} a_{ij} \boldsymbol{e}'_i \tag{A.5}$$

for some a_{ij}. From this, we obtain

$$\boldsymbol{u}' = \sum_{j=1}^{n} u_j f(\boldsymbol{e}_j) = \sum_{j=1}^{n} u_j \sum_{i=1}^{m} a_{ij} \boldsymbol{e}'_i = \sum_{i=1}^{m} \Big(\sum_{j=1}^{n} a_{ij} u_j \Big) \boldsymbol{e}'_i. \tag{A.6}$$

Comparing this with Eq. (A.3), we see that

$$u'_i = \sum_{j=1}^{n} a_{ij} u_j. \tag{A.7}$$

This means that the vector $\boldsymbol{u}' = \big(u'_i\big)$ is obtained by multiplying the vector $\boldsymbol{u} = \big(u_i\big)$ by an $m \times n$ matrix whose (i, j) element is a_{ij}, which we abbreviate to $\big(a_{ij}\big)$, i.e.,

$$\begin{pmatrix} u'_1 \\ \vdots \\ u'_m \end{pmatrix} = \begin{pmatrix} a_{11} & \cdots & a_{1n} \\ \vdots & \ddots & \vdots \\ a_{m1} & \cdots & a_{mn} \end{pmatrix} \begin{pmatrix} u_1 \\ \vdots \\ u_n \end{pmatrix}. \tag{A.8}$$

Thus, we conclude that *a linear mapping from $\mathcal{R}^n$ to $\mathcal{R}^m$ is represented by multiplication of an $m \times n$ matrix $\boldsymbol{A} = \big(a_{ij}\big)$.*

A.2 INNER PRODUCT AND NORM

The *inner product* $\langle \boldsymbol{a}, \boldsymbol{b} \rangle$ of vectors $\boldsymbol{a} = \big(a_i\big)$ and $\boldsymbol{b} = \big(b_i\big)$ is defined by

$$\langle \boldsymbol{a}, \boldsymbol{b} \rangle = \boldsymbol{a}^\top \boldsymbol{b} = \sum_{i=1}^{n} a_i b_i. \tag{A.9}$$

This has the following properties.

symmetry: $\langle \boldsymbol{a}, \boldsymbol{b} \rangle = \langle \boldsymbol{b}, \boldsymbol{a} \rangle$.

linearity: $\langle \boldsymbol{a}, \alpha \boldsymbol{b} + \beta \boldsymbol{c} \rangle = \alpha \langle \boldsymbol{a}, \boldsymbol{b} \rangle + \beta \langle \boldsymbol{a}, \boldsymbol{c} \rangle$, where α and β are arbitrary real numbers.

positivity: $\langle \boldsymbol{a}, \boldsymbol{a} \rangle \geq 0$, where equality holds only for $\boldsymbol{a} = \boldsymbol{0}$.

The *norm* $\|\boldsymbol{a}\|$ of vector $\boldsymbol{a} = \left(a_i\right)$ is defined by

$$\|\boldsymbol{a}\| = \sqrt{\langle \boldsymbol{a}, \boldsymbol{a} \rangle} = \sqrt{\sum_{i=1}^{n} a_i^2}. \tag{A.10}$$

A vector with unit norm is said to be a *unit vector*. The norm has the following properties.

positivity: $\|\boldsymbol{a}\| \geq 0$, where the equality holds only for $\boldsymbol{a} = \boldsymbol{0}$.

Schwarz inequality: $-\|\boldsymbol{a}\| \cdot \|\boldsymbol{b}\| \leq \langle \boldsymbol{a}, \boldsymbol{b} \rangle \leq \|\boldsymbol{a}\| \cdot \|\boldsymbol{b}\|$.

triangle inequality: $|\|\boldsymbol{a}\| - \|\boldsymbol{b}\|| \leq \|\boldsymbol{a} + \boldsymbol{b}\| \leq \|\boldsymbol{a}\| + \|\boldsymbol{b}\|$.

Vectors $\boldsymbol{a}$ and $\boldsymbol{b}$ are said to be *orthogonal* if $\langle \boldsymbol{a}, \boldsymbol{b} \rangle = 0$. The triangle inequality is obtained by applying the Schwartz inequality[1] to

$$\|\boldsymbol{a} + \boldsymbol{b}\|^2 = \langle \boldsymbol{a} + \boldsymbol{b}, \boldsymbol{a} + \boldsymbol{b} \rangle = \|\boldsymbol{a}\|^2 + 2\langle \boldsymbol{a}, \boldsymbol{b} \rangle + \|\boldsymbol{b}\|^2 \tag{A.11}$$

and noting that Eq. (A.11) is larger than or equal to $\|\boldsymbol{a}\|^2 - 2\|\boldsymbol{a}\| \cdot \|\boldsymbol{b}\| + \|\boldsymbol{b}\|^2 = (\|\boldsymbol{a}\| - \|\boldsymbol{b}\|)^2$ and smaller than or equal to $\|\boldsymbol{a}\|^2 + 2\|\boldsymbol{a}\| \cdot \|\boldsymbol{b}\| + \|\boldsymbol{b}\|^2 = (\|\boldsymbol{a}\| + \|\boldsymbol{b}\|)^2$. From this, we also see that if $\langle \boldsymbol{a}, \boldsymbol{b} \rangle = 0$, i.e., if $\boldsymbol{a}$ and $\boldsymbol{b}$ are orthogonal, the *Pythagorean theorem*

$$\|\boldsymbol{a} - \boldsymbol{b}\|^2 = \|\boldsymbol{a}\|^2 + \|\boldsymbol{b}\|^2 = \|\boldsymbol{a} + \boldsymbol{b}\|^2, \tag{A.12}$$

holds for mutually orthogonal vectors $\boldsymbol{a}$ and $\boldsymbol{b}$.

A.3 LINEAR FORMS

For a constant vector $\boldsymbol{a} = \left(a_i\right)$ and a variable vector $\boldsymbol{x} = \left(x_i\right)$, the expression

$$L = \langle \boldsymbol{a}, \boldsymbol{x} \rangle = \sum_{i=1}^{n} a_i x_i \tag{A.13}$$

is called a *linear form* in $\boldsymbol{x}$. Differentiating this with respect to x_i, we obtain

$$\frac{\partial L}{\partial x_i} = a_i. \tag{A.14}$$

[1]This term is in honor of the German mathematician Hermann Amandus Schwarz (1843–1921), but this inequality had also been introduced by the French mathematician Augustin-Louis Cauchy (1789–1857) and the Russian mathematician Viktor Bunyakovsky (1804–1889). It is also known, mostly in France (and Russia), as the *Cauchy–(Bunyakowsky)–Schwarz inequality*.

In vector form, it is written as

$$\nabla_{\boldsymbol{x}} \langle \boldsymbol{a}, \boldsymbol{x} \rangle = \boldsymbol{a}, \tag{A.15}$$

where we define the vector operator $\nabla_{\boldsymbol{x}}(\cdots)$ by

$$\nabla_{\boldsymbol{x}}(\cdots) \equiv \begin{pmatrix} \partial(\cdots)/\partial x_1 \\ \vdots \\ \partial(\cdots)/\partial x_n \end{pmatrix}. \tag{A.16}$$

We call this the *gradient* of $\cdots$ and call the symbol ∇ *nabla*.

A.4 QUADRATIC FORMS

For a constant symmetric matrix $\boldsymbol{A} = \left(a_{ij}\right)$ and a variable vector $\boldsymbol{x} = \left(x_i\right)$, we call the expression

$$Q = \langle \boldsymbol{x}, \boldsymbol{A}\boldsymbol{x} \rangle = \sum_{i,j=1}^{n} a_{ij} x_i x_j \tag{A.17}$$

a *quadratic form* in $\boldsymbol{x}$. The reason for restricting $\boldsymbol{A}$ to be a symmetric matrix is as follows. A general square matrix $\boldsymbol{A}$ is decomposed into the sum of its *symmetric part* $\boldsymbol{A}^{(s)}$ and *anti-symmetric part* $\boldsymbol{A}^{(a)}$ in the form

$$\boldsymbol{A} = \boldsymbol{A}^{(s)} + \boldsymbol{A}^{(a)}, \qquad \boldsymbol{A}^{(s)} \equiv \frac{1}{2}(\boldsymbol{A} + \boldsymbol{A}^\top), \quad \boldsymbol{A}^{(a)} \equiv \frac{1}{2}(\boldsymbol{A} - \boldsymbol{A}^\top). \tag{A.18}$$

By definition, $\boldsymbol{A}^{(s)}$ and $\boldsymbol{A}^{(a)}$ are, respectively, a symmetric matrix and an anti-symmetric matrix:

$$\boldsymbol{A}^{(s)\top} = \boldsymbol{A}^{(s)}, \qquad \boldsymbol{A}^{(a)\top} = -\boldsymbol{A}^{(a)}. \tag{A.19}$$

If the matrix $\boldsymbol{A}$ in Eq. (A.17) is not symmetric, we substitute Eq. (A.18) and obtain

$$\langle \boldsymbol{x}, \boldsymbol{A}\boldsymbol{x} \rangle = \langle \boldsymbol{x}, (\boldsymbol{A}^{(s)} + \boldsymbol{A}^{(a)})\boldsymbol{x} \rangle = \langle \boldsymbol{x}, \boldsymbol{A}^{(s)}\boldsymbol{x} \rangle. \tag{A.20}$$

Hence, only the symmetric part of $\boldsymbol{A}$ has a meaning. The reason for $\langle \boldsymbol{x}, \boldsymbol{A}^{(a)}\boldsymbol{x} \rangle = \sum_{i,j=1}^{n} a_{ij}^{(a)} x_i x_j$ being 0 is that for each pair (i, j) the term $a_{ij}^{(a)} x_i x_j$ and the term $a_{ji}^{(a)} x_j x_i (= -a_{ij}^{(a)} x_i x_j)$ cancel each other. Note that $a_{ii}^{(a)} = 0$ for an anti-symmetric matrix by definition.

This observation implies that for an arbitrary $\boldsymbol{x}$, the equality $\langle \boldsymbol{x}, \boldsymbol{A}\boldsymbol{x} \rangle = 0$ does not mean $\boldsymbol{A} = \boldsymbol{O}$. It only means that

$$\text{If } \langle \boldsymbol{x}, \boldsymbol{A}\boldsymbol{x} \rangle = 0 \text{ for arbitrary } \boldsymbol{x}, \text{ then } \boldsymbol{A}^{(s)} = \boldsymbol{O}. \tag{A.21}$$

Similarly, the equality $\langle \boldsymbol{x}, \boldsymbol{A}\boldsymbol{x} \rangle = \langle \boldsymbol{x}, \boldsymbol{B}\boldsymbol{x} \rangle$ for an arbitrary $\boldsymbol{x}$ does not mean $\boldsymbol{A} = \boldsymbol{B}$. It only means that

$$\text{If } \langle \boldsymbol{x}, \boldsymbol{A}\boldsymbol{x} \rangle = \langle \boldsymbol{x}, \boldsymbol{B}\boldsymbol{x} \rangle \text{ for arbitrary } \boldsymbol{x}, \text{ then } \boldsymbol{A}^{(s)} = \boldsymbol{B}^{(s)}. \tag{A.22}$$

From this consideration, we can assume that the matrix $\boldsymbol{A}$ of a quadratic form is symmetric from the beginning. If $\boldsymbol{A}$ is symmetric, then $a_{ij} = a_{ji}$, and hence the terms of Eq. (A.18) that include x_1 are

$$a_{11}x_1^2 + \sum_{j=2}^{n} a_{1j}x_1x_j + \sum_{2=1}^{n} a_{i1}x_ix_1 = a_{11}x_1^2 + 2(a_{12}x_2 + a_{13}x_3 + \cdots + a_{1n}x_n)x_1. \quad (A.23)$$

Differentiating this with respect to x_1, we obtain $2a_{11}x_1 + 2(a_{12}x_2 + a_{13}x_3 + \cdots + a_{1n}x_n) = 2\sum_{j=1}^{n} a_{1j}x_j$. Similar results hold for x_2, ..., x_n, so we obtain

$$\frac{\partial Q}{\partial x_i} = 2\sum_{j=1}^{n} a_{ij}x_j. \quad (A.24)$$

Using the symbol ∇, we rewrite this in vector form as

$$\nabla_{\boldsymbol{x}}\langle \boldsymbol{x}, \boldsymbol{A}\boldsymbol{x}\rangle = 2\boldsymbol{A}\boldsymbol{x}. \quad (A.25)$$

We see that Eqs. (A.15) and (A.25) are, respectively, extensions of the formulas $d(ax)/dx = a$ and $d(Ax^2)/dx = 2Ax$ for one variable to n variables.

A.5 BILINEAR FORMS

For a constant matrix $\boldsymbol{A} = \left(a_{ij}\right)$ and variable vectors $\boldsymbol{x} = \left(x_i\right)$ and $\boldsymbol{y} = \left(y_i\right)$, we call the expression

$$B = \langle \boldsymbol{x}, \boldsymbol{A}\boldsymbol{y}\rangle = \sum_{i,j=1}^{n} a_{ij}x_iy_j \quad (A.26)$$

a *bilinear form* in $\boldsymbol{x}$ and $\boldsymbol{y}$, for which we observe the fundamental equality

$$\langle \boldsymbol{x}, \boldsymbol{A}\boldsymbol{y}\rangle = \langle \boldsymbol{A}^\top\boldsymbol{x}, \boldsymbol{y}\rangle. \quad (A.27)$$

In fact, both sides are equal to $\sum_{i,j=1}^{N} a_{ij}x_iy_j$ from the rule of the product of a matrix and a vector. In contrast to the case of quadratic forms, we see:

$$\text{If } \langle \boldsymbol{x}, \boldsymbol{A}\boldsymbol{y}\rangle = 0 \text{ for arbitrary } \boldsymbol{x} \text{ and } \boldsymbol{y}, \text{ then } \boldsymbol{A} = \boldsymbol{O}. \quad (A.28)$$

$$\text{If } \langle \boldsymbol{x}, \boldsymbol{A}\boldsymbol{y}\rangle = \langle \boldsymbol{x}, \boldsymbol{B}\boldsymbol{y}\rangle \text{ for arbitrary } \boldsymbol{x} \text{ and } \boldsymbol{y}, \text{ then } \boldsymbol{A} = \boldsymbol{B}. \quad (A.29)$$

From Eqs. (A.15) and (A.27), the following identities hold:

$$\nabla_{\boldsymbol{x}}\langle \boldsymbol{x}, \boldsymbol{A}\boldsymbol{y}\rangle = \boldsymbol{A}\boldsymbol{y}, \qquad \nabla_{\boldsymbol{y}}\langle \boldsymbol{x}, \boldsymbol{A}\boldsymbol{y}\rangle = \boldsymbol{A}^\top\boldsymbol{x}. \quad (A.30)$$

A.6 BASIS AND EXPANSION

A set of vectors $\boldsymbol{u}_1$, ..., $\boldsymbol{u}_r$ is said to be an *orthonormal system* if they are all unit vectors and orthogonal to each other, i.e.,

$$\langle \boldsymbol{u}_i, \boldsymbol{u}_j \rangle = \delta_{ij}, \tag{A.31}$$

where δ_{ij} is the *Kronecker delta* (the symbol that takes 1 for $i = j$ and 0 for $i \neq j$).

If an arbitrary vector $\boldsymbol{x}$ can be uniquely expressed as a linear combination of some n vectors $\boldsymbol{u}_1$, ..., $\boldsymbol{u}_n$, they are said to be a *basis* of that space and n is called the *dimension* of that space. An orthonormal system of n vectors $\{\boldsymbol{u}_1, ..., \boldsymbol{u}_n\}$ constitute a basis of the n-dimensional space $\mathcal{R}^n$, called an *orthonormal basis*.

Expressing a given vector $\boldsymbol{x}$ as a linear combination of an orthonormal basis $\{\boldsymbol{u}_i\}$, $i = 1$, ..., n, in the form of

$$\boldsymbol{x} = c_1 \boldsymbol{u}_1 + \cdots + c_n \boldsymbol{u}_n \tag{A.32}$$

is called *expansion* of x in terms of $\{\boldsymbol{u}_i\}$. The square norm of Eq. (A.32) is written as

$$\begin{aligned} \|\boldsymbol{x}\|^2 &= \langle \sum_{i=1}^n c_i \boldsymbol{u}_i, \sum_{j=1}^n c_j \boldsymbol{u}_j \rangle = \sum_{i,j=1}^n c_i c_j \langle \boldsymbol{u}_i, \boldsymbol{u}_j \rangle = \sum_{i,j=1}^n \delta_{ij} c_i c_j = \sum_{i=1}^n c_i^2 \\ &= c_1^2 + \cdots + c_n^2, \end{aligned} \tag{A.33}$$

where we write $\langle \sum_{i=1}^n c_i \boldsymbol{u}_i, \sum_{j=1}^n c_j \boldsymbol{u}_j \rangle$ instead of $\langle \sum_{i=1}^n c_i \boldsymbol{u}_i, \sum_{i=1}^n c_i \boldsymbol{u}_i \rangle$ for avoiding the confusion of running indices in the summation. Note that if the Kronecker delta δ_{ij} appears in a summation $\sum$ with respect to i or j or both, only terms for which $i = j$ survive.

Computing the inner product of $\boldsymbol{u}_i$ and Eq. (A.32) and noting that $\{\boldsymbol{u}_i\}$ is an orthonormal system, we obtain

$$\langle \boldsymbol{u}_i, \boldsymbol{x} \rangle = c_1 \langle \boldsymbol{u}_i, \boldsymbol{u}_1 \rangle + \cdots + c_i \langle \boldsymbol{u}_i, \boldsymbol{u}_i \rangle + \cdots + c_n \langle \boldsymbol{u}_i, \boldsymbol{u}_n \rangle = c_i. \tag{A.34}$$

Hence, the expansion of Eq. (A.32) is written in the form

$$\boldsymbol{x} = \langle \boldsymbol{u}_1, \boldsymbol{x} \rangle \boldsymbol{u}_1 + \cdots + \langle \boldsymbol{u}_n, \boldsymbol{x} \rangle \boldsymbol{u}_n. \tag{A.35}$$

Since $\{\boldsymbol{u}_i\}$ is a basis, the expansion expression is unique. From Eq. (A.33), its square norm is written as

$$\|\boldsymbol{x}\|^2 = \langle \boldsymbol{u}_1, \boldsymbol{x} \rangle^2 + \cdots + \langle \boldsymbol{u}_n, \boldsymbol{x} \rangle^2. \tag{A.36}$$

A.7 LEAST-SQUARES APPROXIMATION

For an orthonormal system $\{\boldsymbol{u}_i\}$, $i = 1$, ..., r, of r ($\leq n$) vectors, vector $\boldsymbol{x}$ is not necessarily expandable in the form of Eq. (A.32). However, we can determine the expansion coefficients c_i, $i = 1$, ..., r, in such a way that

$$J = \|\boldsymbol{x} - (c_1 \boldsymbol{u}_1 + \cdots + c_r \boldsymbol{u}_r)\|^2 \tag{A.37}$$

is minimized. Expansion in this way is called the *least-squares approximation*. The above expression is rewritten as

$$J = \langle \boldsymbol{x} - \sum_{i=1}^{r} c_i \boldsymbol{u}_i, \boldsymbol{x} - \sum_{j=1}^{r} c_j \boldsymbol{u}_j \rangle. \tag{A.38}$$

Differentiating this with respect to c_k, we obtain

$$\begin{aligned} \frac{\partial J}{\partial c_k} &= \langle -\boldsymbol{u}_k, \boldsymbol{x} - \sum_{j=1}^{r} c_j \boldsymbol{u}_j \rangle + \langle \boldsymbol{x} - \sum_{i=1}^{r} c_i \boldsymbol{u}_i, -\boldsymbol{u}_k \rangle \\ &= -2\langle \boldsymbol{u}_k, \boldsymbol{x} \rangle + 2\sum_{j=1}^{r} c_j \langle \boldsymbol{u}_k, \boldsymbol{u}_j \rangle = -2\langle \boldsymbol{u}_k, \boldsymbol{x} \rangle + 2c_k. \end{aligned} \tag{A.39}$$

Equating this to 0, we obtain $c_k = \langle \boldsymbol{u}_k, \boldsymbol{x} \rangle$. Hence, the least-squares expansion has the form

$$\boldsymbol{x} \approx \langle \boldsymbol{u}_1, \boldsymbol{x} \rangle \boldsymbol{u}_1 + \cdots + \langle \boldsymbol{u}_r, \boldsymbol{x} \rangle \boldsymbol{u}_r. \tag{A.40}$$

This corresponds to *truncating the expansion of Eq. (A.35) up to the r th term*.

The set $\mathcal{U}$ of all vectors that can be expressed as linear combinations of $\boldsymbol{u}_1$, ..., $\boldsymbol{u}_r$ is called the *subspace spanned* by $\boldsymbol{u}_1$, ..., $\boldsymbol{u}_r$. If $\boldsymbol{u}_1$, ..., $\boldsymbol{u}_r$ are an orthonormal system, they form the basis of the subspace $\mathcal{U}$, and its dimension is r. The right side of Eq. (A.40) equals the projection $\boldsymbol{P}_{\mathcal{U}}\boldsymbol{x}$ of $\boldsymbol{x}$ onto the subspace $\mathcal{U}$ by the projection matrix $\boldsymbol{P}_{\mathcal{U}} = \boldsymbol{u}_1\boldsymbol{u}_1^\top + \cdots + \boldsymbol{u}_r\boldsymbol{u}_r^\top$ (↪ Section 2.3). Namely, *least-squares approximation equals projection onto a subspace*. From this observation, we see that if the vector $\boldsymbol{x}$ is in the subspace $\mathcal{U}$, Eq. (A.40) holds with equality, i.e.,

$$\boldsymbol{x} = \langle \boldsymbol{u}_1, \boldsymbol{x} \rangle \boldsymbol{u}_1 + \cdots + \langle \boldsymbol{u}_r, \boldsymbol{x} \rangle \boldsymbol{u}_r \quad (= \boldsymbol{P}_{\mathcal{U}}\boldsymbol{x}). \tag{A.41}$$

The vectors $\{\boldsymbol{u}_i\}$, $i = 1$, ..., r, are now an orthonormal basis of $\mathcal{U}$, and hence this expansion expression is unique. Its square norm is given by

$$\|\boldsymbol{x}\|^2 = \langle \boldsymbol{u}_1, \boldsymbol{x} \rangle^2 + \cdots + \langle \boldsymbol{u}_r, \boldsymbol{x} \rangle^2. \tag{A.42}$$

A.8 LAGRANGE'S METHOD OF INDETERMINATE MULTIPLIERS

The maximum/minimum of a function $f(\boldsymbol{x})$ of variable $\boldsymbol{x}$ is computed by solving

$$\nabla_{\boldsymbol{x}} f = \boldsymbol{0}, \tag{A.43}$$

provided no constraint exists on $\boldsymbol{x}$. If $\boldsymbol{x}$ is constrained to be such that

$$g(\boldsymbol{x}) = 0, \tag{A.44}$$

we introduce a *Lagrange multiplier* λ and consider

$$F = f(\boldsymbol{x}) - \lambda g(\boldsymbol{x}). \tag{A.45}$$

Differentiating this with respect to $\boldsymbol{x}$ and letting the result be $\mathbf{0}$, we obtain

$$\nabla_{\boldsymbol{x}} f - \lambda \nabla_{\boldsymbol{x}} g = \mathbf{0}. \tag{A.46}$$

Solving this together with Eq. (A.44), we can determine $\boldsymbol{x}$ and λ. This process is called *Lagrange's method of indeterminate multipliers*.

We should note that what we obtain by this method is in general an extreme value; a maximum, a minimum, or other types of extremum including an inflection points. However, if it is known from the properties of the problem that it has a unique maximum or a unique minimum, this method is a very convenient and practical means of computing it.

If multiple constraints exist in the form

$$g_1(\boldsymbol{x}) = 0, \qquad \ldots, \qquad g_m(\boldsymbol{x}) = 0, \tag{A.47}$$

we introduce Lagrange multipliers λ_1, ..., λ_m corresponding to individual constraints and consider

$$F = f(\boldsymbol{x}) - \lambda_1 g_1(\boldsymbol{x}) - \cdots - \lambda_m g_m(\boldsymbol{x}) = f(\boldsymbol{x}) - \langle \boldsymbol{\lambda}, \boldsymbol{g}(\boldsymbol{x}) \rangle, \tag{A.48}$$

where we write the m Lagrange multipliers for the m constraints of Eq. (A.47) in the vector form

$$\boldsymbol{\lambda} = \begin{pmatrix} \lambda_1 \\ \vdots \\ \lambda_m \end{pmatrix}, \qquad \boldsymbol{g}(\boldsymbol{x}) = \begin{pmatrix} g_1(\boldsymbol{x}) \\ \vdots \\ g_m(\boldsymbol{x}) \end{pmatrix}. \tag{A.49}$$

Differentiating Eq. (A.48) with respect to $\boldsymbol{x}$ and letting the result be $\mathbf{0}$, we obtain

$$\nabla_{\boldsymbol{x}} f - \langle \boldsymbol{\lambda}, \nabla_{\boldsymbol{x}} \boldsymbol{g}(\boldsymbol{x}) \rangle = \mathbf{0}. \tag{A.50}$$

Solving this together with Eq. (A.47), we can determine $\boldsymbol{x}$ and λ_1, ..., λ_m.

A.9 EIGENVALUES AND EIGENVECTORS

For an $n \times n$ symmetric matrix $\boldsymbol{A}$, we call the value λ that satisfies

$$\boldsymbol{A}\boldsymbol{u} = \lambda \boldsymbol{u}, \qquad \boldsymbol{u} \neq \mathbf{0} \tag{A.51}$$

an *eigenvalue* of $\boldsymbol{A}$. The vector $\boldsymbol{u}$ ($\neq \mathbf{0}$) is called the corresponding *eigenvector*. Equation (A.51) is rewritten in the form

$$(\lambda \boldsymbol{I} - \boldsymbol{A})\boldsymbol{u} = \mathbf{0}. \tag{A.52}$$

This is a set of linear equations in $\boldsymbol{u}$. As is well known, this has a solution $\boldsymbol{u} \neq \boldsymbol{0}$ if and only if the determinant of the coefficient matrix is 0, i.e.,

$$\phi(\lambda) = |\lambda \boldsymbol{I} - \boldsymbol{A}| = 0. \tag{A.53}$$

This is called the *characteristic equation*, where $|\cdot|$ denotes the determinant; $\phi(\lambda)$ is an nth degree polynomial in λ, called the *characteristic polynomial*. Since Eq. (A.53) is an nth order algebraic equation with real coefficients, it has in general n solutions in the complex domain. Hence, matrix $\boldsymbol{A}$ has n eigenvalues and n corresponding eigenvectors. For a symmetric matrix, however, *all eigenvalues are real and the corresponding eigenvectors consist of real components*. This can be shown as follows.

Let λ be an eigenvalue (possibly a complex number) of $\boldsymbol{A}$, and $\boldsymbol{u}$ the corresponding eigenvector (possibly with complex components). Their defining equation and its complex conjugate on both sides are written as

$$\boldsymbol{A}\boldsymbol{u} = \lambda \boldsymbol{u}, \qquad \boldsymbol{A}\bar{\boldsymbol{u}} = \bar{\lambda}\bar{\boldsymbol{u}}, \tag{A.54}$$

where bar designates complex conjugate. The inner product of the first equation and $\bar{\boldsymbol{u}}$ on both sides and the inner product of the second equation and $\boldsymbol{u}$ on both sides are, respectively,

$$\langle \bar{\boldsymbol{u}}, \boldsymbol{A}\boldsymbol{u} \rangle = \lambda \langle \bar{\boldsymbol{u}}, \boldsymbol{u} \rangle, \qquad \langle \boldsymbol{u}, \boldsymbol{A}\bar{\boldsymbol{u}} \rangle = \bar{\lambda} \langle \boldsymbol{u}, \bar{\boldsymbol{u}} \rangle. \tag{A.55}$$

If we write $\boldsymbol{u} = \left(u_i\right)$ $(\neq \boldsymbol{0})$, we see that

$$\langle \bar{\boldsymbol{u}}, \boldsymbol{u} \rangle = \langle \boldsymbol{u}, \bar{\boldsymbol{u}} \rangle = \sum_{i=1}^{n} \bar{u}_i u_i = \sum_{i=1}^{n} |u_i|^2 > 0, \tag{A.56}$$

where $|\cdot|$ denotes the absolute value of a complex number. Since $\boldsymbol{A}$ is symmetric, we see from Eq. (A.27)

$$\langle \boldsymbol{u}, \boldsymbol{A}\bar{\boldsymbol{u}} \rangle = \langle \boldsymbol{A}^\top \boldsymbol{u}, \bar{\boldsymbol{u}} \rangle = \langle \boldsymbol{A}\boldsymbol{u}, \bar{\boldsymbol{u}} \rangle = \langle \bar{\boldsymbol{u}}, \boldsymbol{A}\boldsymbol{u} \rangle. \tag{A.57}$$

Hence, Eq. (A.55) implies that $\lambda = \bar{\lambda}$ and hence λ is a real number. A set of simultaneous linear equations $\boldsymbol{A}\boldsymbol{u} = \lambda \boldsymbol{u}$ in unknown $\boldsymbol{u}$ for a matrix $\boldsymbol{A}$ of real elements and a real number λ can be solved using arithmetic operations and substitutions, so the resulting solution $\boldsymbol{u}$ also has real components.

Furthermore, *eigenvectors for different eigenvalues are mutually orthogonal*. This is shown as follows. Let $\boldsymbol{u}$ and $\boldsymbol{u}'$ be the eigenvectors of $\boldsymbol{A}$ for eigenvalues λ and λ' $(\lambda \neq \lambda')$, and write

$$\boldsymbol{A}\boldsymbol{u} = \lambda \boldsymbol{u}, \qquad \boldsymbol{A}\boldsymbol{u}' = \lambda' \boldsymbol{u}'. \tag{A.58}$$

The inner product of the first equation and $\boldsymbol{u}'$ on both sides and the inner product of the second equation with $\boldsymbol{u}$ on both sides are

$$\langle \boldsymbol{u}', \boldsymbol{A}\boldsymbol{u} \rangle = \lambda \langle \boldsymbol{u}', \boldsymbol{u} \rangle, \qquad \langle \boldsymbol{u}, \boldsymbol{A}\boldsymbol{u}' \rangle = \lambda' \langle \boldsymbol{u}, \boldsymbol{u}' \rangle. \tag{A.59}$$

Since $\boldsymbol{A}$ is symmetric, we see from Eq. (A.27) that

$$\langle \boldsymbol{u}', \boldsymbol{A}\boldsymbol{u}\rangle = \langle \boldsymbol{A}^\top \boldsymbol{u}', \boldsymbol{u}\rangle = \langle \boldsymbol{A}\boldsymbol{u}', \boldsymbol{u}\rangle = \langle \boldsymbol{u}, \boldsymbol{A}\boldsymbol{u}'\rangle. \tag{A.60}$$

Hence, from Eq. (A.59), we obtain

$$\lambda\langle \boldsymbol{u}', \boldsymbol{u}\rangle = \lambda'\langle \boldsymbol{u}, \boldsymbol{u}'\rangle \quad (= \lambda'\langle \boldsymbol{u}', \boldsymbol{u}\rangle), \tag{A.61}$$

which means $(\lambda - \lambda')\langle \boldsymbol{u}', \boldsymbol{u}\rangle = 0$. Since $\lambda \neq \lambda'$, this implies $\langle \boldsymbol{u}', \boldsymbol{u}\rangle = 0$, i.e., $\boldsymbol{u}$ and $\boldsymbol{u}'$ are orthogonal to each other.

If there is multiplicity among the n eigenvalues, the eigenvectors for multiple eigenvalue are not unique. However, their arbitrary linear combinations are also eigenvalues for the same eigenvalue. Hence, we can choose them, using, e.g., the Schmidt orthogonalization ($\hookrightarrow$ Section 2.5), to be mutually orthogonal vectors. As can be seen from Eq. (A.51), if $\boldsymbol{u}$ is an eigenvector, its arbitrary constant multiple $c\boldsymbol{u}$ ($c \neq 0$) is also an eigenvector for the same eigenvalue. As a result, the eigenvectors $\{\boldsymbol{u}_i\}$, $i = 1, ..., n$, of a symmetric matrix can be chosen to be an orthonormal system of vectors.

A.10 MAXIMUM AND MINIMUM OF A QUADRATIC FORM

Consider a quadratic form $\langle \boldsymbol{v}, \boldsymbol{A}\boldsymbol{v}\rangle$ of an $n \times n$ symmetric matrix $\boldsymbol{A}$ in unit vector $\boldsymbol{v}$. Let $\lambda_1 \geq \cdots \geq \lambda_n$ be the n eigenvalues of $\boldsymbol{A}$, and $\{\boldsymbol{u}_i\}$, $i = 1, ..., n$, the corresponding orthonormal system of the corresponding eigenvectors. Since, as shown in Section A.6, an arbitrary unit vector $\boldsymbol{v}$ can be expanded in the form $\boldsymbol{v} = \sum_{i=1}^n c_i \boldsymbol{u}_i$, $\sum_{i=1}^n c_i^2 = 1$, we can rewrite the quadratic form as

$$\begin{aligned}\langle \boldsymbol{v}, \boldsymbol{A}\boldsymbol{v}\rangle &= \langle \sum_{i=1}^n c_i \boldsymbol{u}_i, \boldsymbol{A}\sum_{j=1}^n c_j \boldsymbol{u}_j\rangle = \sum_{i,j=1}^n c_i c_j \langle \boldsymbol{u}_i, \boldsymbol{A}\boldsymbol{u}_j\rangle = \sum_{i,j=1}^n c_i c_j \langle \boldsymbol{u}_i, \lambda_j \boldsymbol{u}_j\rangle \\ &= \sum_{i,j=1}^n c_i c_j \lambda_j \langle \boldsymbol{u}_i, \boldsymbol{u}_j\rangle = \sum_{i,j=1}^n c_i c_j \lambda_j \delta_{ij} = \sum_{i=1}^n c_i^2 \lambda_i \leq \lambda_1 \sum_{i=2}^n c_i^2 = \lambda_1,\end{aligned} \tag{A.62}$$

where equality holds when $c_1 = 1$ and $c_2 = \cdots = c_n = 0$, i.e., when $\boldsymbol{v} = \boldsymbol{u}_1$. Reversing the inequality, we similarly obtain

$$\langle \boldsymbol{v}, \boldsymbol{A}\boldsymbol{v}\rangle = \sum_{i=1}^n c_i^2 \lambda_i \geq \lambda_n \sum_{i=2}^n c_i^2 = \lambda_n. \tag{A.63}$$

Thus, we conclude that *the maximum and minimum of a quadratic form $\langle \boldsymbol{v}, \boldsymbol{A}\boldsymbol{v}\rangle$ of a symmetric matrix $\boldsymbol{A}$ in unit vector $\boldsymbol{v}$ equal the maximum eigenvalue λ_1 and the minimum eigenvalue λ_n of $\boldsymbol{A}$, respectively, $\boldsymbol{v}$ being equal to the corresponding unit eigenvectors $\boldsymbol{u}_1$ and $\boldsymbol{u}_n$, respectively.*

Next, consider a unit vector $\boldsymbol{v}$ orthogonal to $\boldsymbol{u}_1$. An arbitrary unit vector $\boldsymbol{v}$ orthogonal to $\boldsymbol{u}_1$ can be expanded in the form $\boldsymbol{v} = \sum_{i=2}^n c_i \boldsymbol{u}_i$, $\sum_{i=2}^n c_i^2 = 1$. Hence, the quadratic form

$\langle \boldsymbol{v}, \boldsymbol{A}\boldsymbol{v} \rangle$ can be written as

$$\begin{aligned}
\langle \boldsymbol{v}, \boldsymbol{A}\boldsymbol{v} \rangle &= \langle \sum_{i=2}^{n} c_i \boldsymbol{u}_i, \boldsymbol{A} \sum_{j=2}^{n} c_j \boldsymbol{u}_j \rangle = \sum_{i,j=2}^{n} c_i c_j \langle \boldsymbol{u}_i, \boldsymbol{A}\boldsymbol{u}_j \rangle = \sum_{i,j=2}^{n} c_i c_j \langle \boldsymbol{u}_i, \lambda_j \boldsymbol{u}_j \rangle \\
&= \sum_{i,j=2}^{n} c_i c_j \lambda_j \langle \boldsymbol{u}_i, \boldsymbol{u}_j \rangle = \sum_{i,j=2}^{n} c_i c_j \lambda_j \delta_{ij} = \sum_{i=2}^{n} c_i^2 \lambda_i \leq \lambda_2 \sum_{i=2}^{n} c_i^2 = \lambda_2, \quad \text{(A.64)}
\end{aligned}$$

where equality holds when $c_2 = 1$ and $c_3 = \cdots = c_n = 0$, i.e., when $\boldsymbol{v} = \boldsymbol{u}_2$. Hence, $\langle \boldsymbol{v}, \boldsymbol{A}\boldsymbol{v} \rangle$ *for a unit vector* $\boldsymbol{v}$ *orthogonal to* $\boldsymbol{u}_1$ *takes its maximum when* $\boldsymbol{v}$ *equals the unit eigenvector* $\boldsymbol{u}_2$ *for the second largest eigenvalue* λ_2*, the maximum value being* λ_2.

By the same argument, we conclude that $\langle \boldsymbol{v}, \boldsymbol{A}\boldsymbol{v} \rangle$ *for a unit vector* $\boldsymbol{v}$ *orthogonal to* $\boldsymbol{u}_1$, ..., $\boldsymbol{u}_{m-1}$ *takes its maximum when* $\boldsymbol{v}$ *equals the unit eigenvector* $\boldsymbol{u}_m$ *for the* m*th largest eigenvalue* λ_m*, the maximum being* λ_m. The same holds for the minimum: $\langle \boldsymbol{v}, \boldsymbol{A}\boldsymbol{v} \rangle$ *for a unit vector* $\boldsymbol{v}$ *orthogonal to* $\boldsymbol{u}_{n-m}$, ..., $\boldsymbol{u}_n$ *takes its minimum when* $\boldsymbol{v}$ *equals the unit eigenvector* $\boldsymbol{u}_{n-m+1}$ *for the* m*th smallest eigenvalue* λ_{n-m+1}*, the maximum being* λ_{n-m+}.

APPENDIX B

Answers

Chapter 2

2.1. (1) From the definition of matrix multiplication comes the following identity:

$$\begin{pmatrix} a_1 \\ \vdots \\ a_m \end{pmatrix} \begin{pmatrix} b_1 & \cdots & b_n \end{pmatrix} = \begin{pmatrix} a_1 b_1 & \cdots & a_1 b_n \\ \vdots & \ddots & \vdots \\ a_m b_1 & \cdots & a_m b_n \end{pmatrix}.$$

(2) The trace of the above matrix is $\sum_{i=1}^{n} a_i b_i = \langle \boldsymbol{a}, \boldsymbol{b} \rangle$.

2.2. Let $\{\boldsymbol{u}_i\}$, $i = 1, ..., r$, be the orthonormal basis of $\mathcal{U}$, and express $\overrightarrow{OQ} \in \mathcal{U}$ as $\overrightarrow{OQ} = \sum_{i=1}^{r} c_i \boldsymbol{u}_i$ in terms of the basis vectors. Then,

$$\|\overrightarrow{PQ}\|^2 = \|\overrightarrow{OQ} - \overrightarrow{OP}\|^2 = \|\sum_{i=1}^{r} c_i \boldsymbol{u}_i - \overrightarrow{OP}\|^2 = \langle \sum_{i=1}^{r} c_i \boldsymbol{u}_i - \overrightarrow{OP}, \sum_{j=1}^{r} c_j \boldsymbol{u}_j - \overrightarrow{OP} \rangle.$$

Differentiation of this with respect to c_i is

$$\frac{\partial \|\overrightarrow{PQ}\|^2}{\partial c_i} = 2\langle \boldsymbol{u}_i, \sum_{j=1}^{r} c_j \boldsymbol{u}_j - \overrightarrow{OP} \rangle = 2\langle \boldsymbol{u}_i, \overrightarrow{PQ} \rangle.$$

Since this vanishes at the closest point to P, we see that $\overrightarrow{PQ}$ is orthogonal to all the basis vectors $\{\boldsymbol{u}_i\}$, $i = 1, ..., r$, of $\mathcal{U}$. Hence, $\overrightarrow{OQ}$ is the projection of $\overrightarrow{OP}$.

2.3. They are shown as follows:

$$\boldsymbol{P}_{\mathcal{U}}^\top = \Big(\sum_{i=1}^{r} \boldsymbol{u}_i \boldsymbol{u}_i^\top\Big)^\top = \sum_{i=1}^{r} \boldsymbol{u}_i \boldsymbol{u}_i^\top = \boldsymbol{P}_{\mathcal{U}},$$

$$\begin{aligned} \boldsymbol{P}_{\mathcal{U}}^2 &= \Big(\sum_{i=1}^{r} \boldsymbol{u}_i \boldsymbol{u}_i^\top\Big)\Big(\sum_{j=1}^{r} \boldsymbol{u}_j \boldsymbol{u}_j^\top\Big) = \sum_{i,j=1}^{r} \boldsymbol{u}_i \boldsymbol{u}_i^\top \boldsymbol{u}_j \boldsymbol{u}_j^\top = \sum_{i,j=1}^{r} \boldsymbol{u}_i \langle \boldsymbol{u}_i, \boldsymbol{u}_j \rangle \boldsymbol{u}_j^\top \\ &= \sum_{i,j=1}^{r} \delta_{ij} \boldsymbol{u}_i \boldsymbol{u}_j^\top = \sum_{i=1}^{r} \boldsymbol{u}_i \boldsymbol{u}_i^\top = \boldsymbol{P}_{\mathcal{U}}. \end{aligned}$$

2.4. As is well known, an $n \times n$ symmetric matrix $\boldsymbol{P}$ has n real eigenvalues λ_1, ..., λ_n, the corresponding eigenvectors $\boldsymbol{u}_1$, ..., $\boldsymbol{u}_n$ being an orthonormal system. If we multiply $\boldsymbol{P}\boldsymbol{u}_i = \lambda_i \boldsymbol{u}_i$ by $\boldsymbol{P}$ from left on both sides, we have

$$\boldsymbol{P}^2\boldsymbol{u}_i = \lambda_i \boldsymbol{P}\boldsymbol{u}_i = \lambda_i^2 \boldsymbol{u}_i.$$

If $\boldsymbol{P}$ is idempotent, the left side is $\boldsymbol{P}\boldsymbol{u}_i = \lambda_i \boldsymbol{u}_i$. Hence, $\lambda_i = \lambda_i^2$, i.e., $\lambda_i = 0, 1$. Let $\lambda_1 = \cdots = \lambda_r = 1, \lambda_{r+1} = \cdots = \lambda_n = 0$. Then,

$$\boldsymbol{P}\boldsymbol{u}_i = \boldsymbol{u}_i, \quad i = 1, ..., r, \qquad \boldsymbol{P}\boldsymbol{u}_i = \boldsymbol{0}, \quad i = r+1, ..., n.$$

From Eq. (2.6), we see that $\boldsymbol{P}$ is the projection matrix onto the subspace spanned by $\boldsymbol{u}_1$, ..., $\boldsymbol{u}_r$.

2.5. If subspaces $\mathcal{U}$ and $\mathcal{V}$ have respective dimensions r and $n - r$ and if $\mathcal{U} \cap \mathcal{V} = \{\boldsymbol{0}\}$, we can choose a basis $\{\boldsymbol{u}_1, ..., \boldsymbol{u}_r\}$ of $\mathcal{U}$ and a basis $\{\boldsymbol{u}_{r+1}, ..., \boldsymbol{u}_n\}$ of $\mathcal{V}$ so that $\{\boldsymbol{u}_1, ..., \boldsymbol{u}_n\}$ spans $\mathcal{R}^n$. Since any vector $\boldsymbol{x} \in \mathcal{R}^n$ is uniquely expressed as a linear combination of $\{\boldsymbol{u}_i\}$, $i = 1$, ..., n, it is written as Eq. (2.22) uniquely.

2.6. (1) The vector product $\boldsymbol{u}_2 \times \boldsymbol{u}_3$ is orthogonal to $\boldsymbol{u}_1$ and $\boldsymbol{u}_2$, so $\boldsymbol{u}_1^\dagger$ is orthogonal to $\boldsymbol{u}_1$ and $\boldsymbol{u}_2$. On the other hand, $\langle \boldsymbol{u}_1^\dagger, \boldsymbol{u}_1 \rangle$ equals 1 due to the definition of the scalar triple product $|\boldsymbol{u}_1, \boldsymbol{u}_2, \boldsymbol{u}_3|$. The same holds for $\boldsymbol{u}_2^\dagger$ and $\boldsymbol{u}_3^\dagger$. Hence, Eq. (2.24) holds.

(2) The definition of Eq, (2.31) implies

$$\begin{pmatrix} \boldsymbol{u}_1^{\dagger\top} \\ \vdots \\ \boldsymbol{u}_n^{\dagger\top} \end{pmatrix} \boldsymbol{U} = \begin{pmatrix} \boldsymbol{u}_1^{\dagger\top} \\ \vdots \\ \boldsymbol{u}_n^{\dagger\top} \end{pmatrix} \begin{pmatrix} \boldsymbol{u}_1 & \cdots & \boldsymbol{u}_n \end{pmatrix} = \boldsymbol{I},$$

which means Eq. (2.24).

2.7. (1) Since the matrix $\boldsymbol{P}$ maps the basis vector $\boldsymbol{e}_i$ to the ith column $\boldsymbol{p}_i$ of $\boldsymbol{P}$, we have $\boldsymbol{p}_i = \boldsymbol{P}\boldsymbol{e}_i$. Hence, $\boldsymbol{P}(\boldsymbol{e}_i - \boldsymbol{p}_i) = \boldsymbol{0}$, i.e, $\{\boldsymbol{e}_i - \boldsymbol{p}_i\}$, $i = 1, ..., n$, are all mapped to $\boldsymbol{0}$ by $\boldsymbol{P}$.

(2) It is easy to see that $\mathcal{V} \cap \ker(\boldsymbol{P}) = \{\boldsymbol{0}\}$. In fact, if $\boldsymbol{x} \in \mathcal{V}$, it is expressed as a linear combination of $\{\boldsymbol{p}_i\}$, i.e., $\boldsymbol{x} = c_1\boldsymbol{p}_1 + \cdots + c_n\boldsymbol{p}_i = \boldsymbol{P}\boldsymbol{c}$, for some $\boldsymbol{c} = \left(c_i\right)$. Hence, $\boldsymbol{P}\boldsymbol{x} = \boldsymbol{P}\boldsymbol{P}\boldsymbol{c} = \boldsymbol{P}\boldsymbol{c} = \boldsymbol{x}$. If $\boldsymbol{x}$ belongs to $\ker(\boldsymbol{P})$, i.e., if $\boldsymbol{P}\boldsymbol{x} = \boldsymbol{0}$, we have $\boldsymbol{P}\boldsymbol{x} = \boldsymbol{x} = \boldsymbol{0}$. On the other hand, $\mathcal{V} \cup \ker(\boldsymbol{P}) = \{\boldsymbol{p}_1, ..., \boldsymbol{p}_n\} \cup \{\boldsymbol{e}_1 - \boldsymbol{p}_1, ..., \boldsymbol{e}_n - \boldsymbol{p}_n\}$ contains $\{\boldsymbol{e}_1, ..., \boldsymbol{e}_n\}$, hence has dimension n. It follows that $\mathcal{R} = \mathcal{V} + \ker(\boldsymbol{P})$ is a direct sum decomposition.

(3) Since $\boldsymbol{P}\boldsymbol{x}$ is a linear combination of $\boldsymbol{p}_1$, ..., $\boldsymbol{p}_n$, it belongs to $\mathcal{V}$. On the other hand, $\boldsymbol{P}(\boldsymbol{I} - \boldsymbol{P})\boldsymbol{x} = (\boldsymbol{P} - \boldsymbol{P}^2)\boldsymbol{x} = (\boldsymbol{P} - \boldsymbol{P})\boldsymbol{x} = \boldsymbol{0}$. Hence, Eq. (2.32) is a direct sum decomposition.

Chapter 3

3.1. Suppose some linear combination of $\boldsymbol{u}_1$, ..., $\boldsymbol{u}_m$, $\boldsymbol{u}_i \neq \mathbf{0}$, $i = 1$, ..., m, is $\mathbf{0}$:

$$c_1\boldsymbol{u}_1 + \cdots + c_m\boldsymbol{u}_m = \mathbf{0}.$$

Compute the inner product of this equation and $\boldsymbol{u}_k$ on both sides. Since $\{\boldsymbol{u}_i\}$, $i = 1$, ..., m, are mutually orthogonal, we see that

$$c_k\langle \boldsymbol{u}_k, \boldsymbol{u}_k\rangle = c_k\|\boldsymbol{u}_k\|^2 = 0,$$

and hence $c_k = 0$. This holds for all $k = 1$, ..., m. Hence, the linear combination of $\boldsymbol{u}_1$, ..., $\boldsymbol{u}_m$ is $\mathbf{0}$ only when all the coefficients are 0. Namely, $\boldsymbol{u}_1$, ..., $\boldsymbol{u}_m$ are linearly independent.

3.2. If we write the jth components of $\boldsymbol{a}_i$ and $\boldsymbol{b}_i$ as a_{ji} and b_{ji}, respectively, the (k,l) element of the matrix $\boldsymbol{a}_i\boldsymbol{b}_i^\top$ is $a_{ki}b_{li}$ (↪ Eq. (2.27)). Hence, the (k,l) element of the matrix on the left side of Eq. (3.29) is $\sum_{i=1}^n a_{ki}b_{li}$. Since $\boldsymbol{A} = \begin{pmatrix} \boldsymbol{a}_1 & \cdots & \boldsymbol{a}_n \end{pmatrix} = \begin{pmatrix} a_{ij} \end{pmatrix}$ and $\boldsymbol{B} = \begin{pmatrix} \boldsymbol{b}_1 & \cdots & \boldsymbol{b}_n \end{pmatrix} = \begin{pmatrix} b_{ij} \end{pmatrix}$, we see that $\sum_{i=1}^n a_{ki}b_{li}$ equals the (k,l) element of $\boldsymbol{A}\boldsymbol{B}^\top$.

3.3. The definition of the matrix $\boldsymbol{U}$ implies that

$$\boldsymbol{U}^\top\boldsymbol{U} = \begin{pmatrix} \boldsymbol{u}_1^\top \\ \vdots \\ \boldsymbol{u}_n^\top \end{pmatrix} \begin{pmatrix} \boldsymbol{u}_1 & \cdots & \boldsymbol{u}_n \end{pmatrix} = \begin{pmatrix} \langle \boldsymbol{u}_1, \boldsymbol{u}_1\rangle & \cdots & \langle \boldsymbol{u}_1, \boldsymbol{u}_n\rangle \\ \vdots & \ddots & \vdots \\ \langle \boldsymbol{u}_n, \boldsymbol{u}_1\rangle & \cdots & \langle \boldsymbol{u}_n, \boldsymbol{u}_n\rangle \end{pmatrix}.$$

This equals $\boldsymbol{I}$ if and only if $\langle \boldsymbol{u}_i, \boldsymbol{u}_j\rangle = \delta_{ij}$, i.e., the columns of $\boldsymbol{U}$ form an orthonormal system.

3.4. Equation (2.11) implies that $\boldsymbol{U}^\top$ is the inverse of $\boldsymbol{U}$, i.e., $\boldsymbol{U}^\top = \boldsymbol{U}^{-1}$. Hence,

$$(\boldsymbol{U}^\top)^\top(\boldsymbol{U}^\top) = \boldsymbol{U}\boldsymbol{U}^\top = \boldsymbol{U}\boldsymbol{U}^{-1} = \boldsymbol{I},$$

which means that $\boldsymbol{U}^\top$ is an orthogonal matrix. It follows that the rows of $\boldsymbol{U}$ are also orthonormal.

3.5. (1) Since the determinant is the same if the matrix is transposed, Eq. (3.9) implies that

$$|\boldsymbol{U}|^2 = |\boldsymbol{U}||\boldsymbol{U}| = |\boldsymbol{U}^\top||\boldsymbol{U}| = |\boldsymbol{U}^\top\boldsymbol{U}| = |\boldsymbol{I}| = 1,$$

meaning that $|\boldsymbol{U}| = \pm 1$.

(2) From Eq. (3.7), we see that

$$|A| = |U| \left| \begin{pmatrix} \lambda_1 & & \\ & \ddots & \\ & & \lambda_n \end{pmatrix} \right| |U^\top| = |U|\lambda_1 \cdots \lambda_n |U| = \lambda_1 \cdots \lambda_n$$

3.6. Their product is as follows:

$$\begin{aligned} A^{-1}A &= \Big(\sum_{i=1}^{n} \frac{u_i u_i^\top}{\lambda_i}\Big)\Big(\sum_{j=1}^{n} \lambda_j u_j u_j^\top\Big) = \sum_{i,j=1}^{n} \frac{\lambda_j}{\lambda_i} u_i u_i^\top u_j u_j^\top = \sum_{i,j=1}^{n} \frac{\lambda_j}{\lambda_i} u_i \langle u_i, u_j \rangle u_j^\top \\ &= \sum_{i,j=1}^{n} \frac{\lambda_j}{\lambda_i} \delta_{ij} u_i u_j^\top = \sum_{i=1}^{n} u_i u_i^\top = I. \end{aligned}$$

3.7. From Eq. (3.17) follows

$$\begin{aligned} (\sqrt{A})^2 &= \Big(\sum_{i=1}^{n} \sqrt{\lambda_i} u_i^\top\Big)\Big(\sum_{j=1}^{n} \sqrt{\lambda_j} u_j^\top\Big) = \sum_{i,j=1}^{n} \sqrt{\lambda_i \lambda_j} u_i u_i^\top u_j u_j^\top \\ &= \sum_{i,j=1}^{n} \sqrt{\lambda_i \lambda_j} u_i \langle u_i, u_j \rangle u_j^\top = \sum_{i,j=1}^{n} \sqrt{\lambda_i \lambda_j} \delta_{ij} u_i u_j^\top = \sum_{i=1}^{n} \lambda_i u_i u_i^\top = A. \end{aligned}$$

Using Eq. (3.9), on the other hand, we see from the first of Eq. (3.18) that

$$\begin{aligned} (\sqrt{A})^2 &= U \begin{pmatrix} \sqrt{\lambda_1} & & \\ & \ddots & \\ & & \sqrt{\lambda_n} \end{pmatrix} U^\top U \begin{pmatrix} \sqrt{\lambda_1} & & \\ & \ddots & \\ & & \sqrt{\lambda_n} \end{pmatrix} U^\top \\ &= U \begin{pmatrix} \sqrt{\lambda_1} & & \\ & \ddots & \\ & & \sqrt{\lambda_n} \end{pmatrix} \begin{pmatrix} \sqrt{\lambda_1} & & \\ & \ddots & \\ & & \sqrt{\lambda_n} \end{pmatrix} U^\top \\ &= U \begin{pmatrix} \lambda_1 & & \\ & \ddots & \\ & & \lambda_n \end{pmatrix} U^\top = A. \end{aligned}$$

3.8. The following identity holds:

$$(A^{-1})^N A^N = A^{-1} \cdots A^{-1} A \cdots A = I.$$

This implies that $(A^{-1})^N$ is the inverse of A^N, i.e., $(A^{-1})^N = (A^N)^{-1}$.

3.9. Suppose $\boldsymbol{u}_1, \ldots, \boldsymbol{u}_m$ are not linearly independent but at most r ($< m$) of them are linearly independent. Adjusting their order, we can assume that $\boldsymbol{u}_1, \ldots, \boldsymbol{u}_r$ are linearly independent and that $\boldsymbol{r}_k$, $r < k \leq m$ are expressed as a linear combination of them in the form

$$\boldsymbol{u}_k = c_1\boldsymbol{u}_1 + \cdots + c_r\boldsymbol{u}_r, \qquad (*)$$

where $c_1, \ldots, c_r$ are some complex numbers. Multiplying this expression by $\boldsymbol{A}$ from left on both sides, we obtain

$$\lambda_k\boldsymbol{u}_k = c_1\lambda_1\boldsymbol{u}_1 + \cdots + c_r\lambda_r\boldsymbol{u}_r.$$

Multiplying ($*$) by λ_k on both sides, we obtain

$$\lambda_k\boldsymbol{u}_k = c_1\lambda_k\boldsymbol{u}_1 + \cdots + c_r\lambda_k\boldsymbol{u}_r.$$

Hence,

$$\boldsymbol{0} = c_1(\lambda_1 - \lambda_k)\boldsymbol{u}_1 + \cdots + c_r(\lambda_r - \lambda_k)\boldsymbol{u}_r.$$

We are assuming that $\boldsymbol{u}_1, \ldots, \boldsymbol{u}_r$ are linearly independent and that all eigenvalues $\lambda_1, \ldots, \lambda_m$ are distinct, so we conclude that $c_1 = \cdots = c_r = 0$, meaning that $\boldsymbol{u}_k$ ($\neq \boldsymbol{0}$) equals $\boldsymbol{0}$, a contradiction.

3.10. From Eq. (3.31) and the first equation of Eq. (3.33), we see that

$$(\boldsymbol{A} - \lambda\boldsymbol{I})\boldsymbol{u}_1 = (\boldsymbol{A} - \lambda\boldsymbol{I})^n\boldsymbol{u}_n = \boldsymbol{0},$$

hence $\boldsymbol{A}\boldsymbol{u}_1 = \lambda\boldsymbol{u}_1$, i.e., $\boldsymbol{u}_1$ is an eigenvector of $\boldsymbol{A}$ for eigenvalue λ. Each equation of Eq. (3.33) is rewritten in the form

$$\begin{aligned}\boldsymbol{u}_1 &= (\boldsymbol{A} - \lambda\boldsymbol{I})\boldsymbol{u}_2,\\ \boldsymbol{u}_2 &= (\boldsymbol{A} - \lambda\boldsymbol{I})\boldsymbol{u}_3,\\ &\cdots\\ \boldsymbol{u}_{n-1} &= (\boldsymbol{A} - \lambda\boldsymbol{I})\boldsymbol{u}_n.\end{aligned}$$

Hence, we obtain

$$\begin{aligned}\boldsymbol{A}\boldsymbol{u}_1 &= \lambda\boldsymbol{u}_1,\\ \boldsymbol{A}\boldsymbol{u}_2 &= \lambda\boldsymbol{u}_2 + \boldsymbol{u}_1,\\ &\cdots\\ \boldsymbol{A}\boldsymbol{u}_n &= \lambda\boldsymbol{u}_n + \boldsymbol{u}_{n-1}.\end{aligned}$$

Rewriting these, we obtain

$$AU = A\begin{pmatrix} u_1 & \cdots & u_n \end{pmatrix} = \begin{pmatrix} u_1 & \cdots & u_n \end{pmatrix}\begin{pmatrix} \lambda & 1 & & & \\ & \lambda & 1 & & \\ & & \ddots & \ddots & \\ & & & & 1 \\ & & & & \lambda \end{pmatrix}$$

$$= U\begin{pmatrix} \lambda & 1 & & & \\ & \lambda & 1 & & \\ & & \ddots & \ddots & \\ & & & & 1 \\ & & & & \lambda \end{pmatrix}.$$

Multiplying U^{-1} from left on both sides, we obtain Eq. (3.34).

Chapter 4

4.1. Evidently, $AA^\top$ and $A^\top A$ are both symmetric matrices. Suppose $AA^\top$ has eigenvalue λ for the eigenvector u ($\neq 0$). Computing the inner product of $AA^\top u = \lambda u$ with u on both sides, we see that

$$\langle u, AA^\top u\rangle = \lambda\langle u, u\rangle = \lambda\|u\|^2.$$

Since

$$\langle u, AA^\top u\rangle = \langle A^\top u, A^\top u\rangle = \|A^\top u\|^2 \geq 0,$$

($\hookrightarrow$ Appendix Eq. (A.27)), we conclude that $\lambda \geq 0$. Similarly, suppose $A^\top A$ has eigenvalue λ' for the eigenvector v ($\neq 0$). Computing the inner product of $A^\top A v = \lambda' v$ with v on both sides, we see that

$$\langle v, A^\top A v\rangle = \lambda'\langle v, v\rangle = \lambda'\|v\|^2,$$

$$\langle v, A^\top A v\rangle = \langle Av, Av\rangle = \|Av\|^2 \geq 0.$$

Hence, $\lambda' \geq 0$.

4.2. As shown in the preceding problem, the eigenvalues of $AA^\top$ are all positive or zero. Since $A \neq O$, at least one of them is positive, say σ^2. Let u be its eigenvector. Multiplying $AA^\top u = \sigma^2 u$ by $A^\top$ on both sides from left, we obtain

$$A^\top AA^\top u = \sigma^2 A^\top u.$$

If we let $v = A^\top u/\sigma$, this equality is written as $A^\top A(\sigma v) = \sigma^3 v$, i.e.,

$$A^\top A v = \sigma^2 v.$$

This implies that $A^\top A$ has eigenvalue σ^2 for the eigenvector v. Conversely, suppose $A^\top A$ has eigenvalue σ^2 for the eigenvector v. Multiplying $A^\top A v = \sigma^2 v$ by A on both sides from left, we obtain

$$AA^\top A v = \sigma^2 A v.$$

If we let $u = Av/\sigma$, this equality is written as $AA^\top(\sigma u) = \sigma^3 u$, i.e.,

$$AA^\top u = \sigma^2 u.$$

This implies that $AA^\top$ has eigenvalue σ^2 for the eigenvector u. Thus, if one of $A^\top A$ and $A^\top A$ has a positive eigenvalue σ^2, it is also an eigenvalue of the other, and their eigenvectors v and u are related by

$$v = \frac{A^\top u}{\sigma}, \qquad u = \frac{Av}{\sigma}.$$

Namely, Eq. (4.1) holds.

4.3. (1) Suppose $AA^\top u = \mathbf{0}$. Computing its inner product with u on both sides, we obtain

$$\langle u, AA^\top u\rangle = \langle A^\top u, A^\top u\rangle = \|A^\top u\|^2 = 0.$$

($\hookrightarrow$ Eq. (A.27)). Hence, $A^\top u = \mathbf{0}$.

(2) Suppose $A^\top A v = \mathbf{0}$. Computing its inner product with v on both sides, we obtain

$$\langle v, A^\top A v\rangle = \langle Av, Av\rangle = \|Av\|^2 = 0.$$

($\hookrightarrow$ Eq. (A.27)). Hence, $Av = \mathbf{0}$.

4.4. Since $\{u_i\}$, $i = 1, ..., r$, are an orthonormal system, the rule of matrix multiplication implies

$$U^\top U = \begin{pmatrix} u_1^\top \\ \vdots \\ u_r^\top \end{pmatrix} \begin{pmatrix} u_1 & \cdots & u_r \end{pmatrix} = \begin{pmatrix} \langle u_1, u_1\rangle & \cdots & \langle u_1, u_r\rangle \\ \vdots & \ddots & \vdots \\ \langle u_r, u_1\rangle & \cdots & \langle u_r, u_r\rangle \end{pmatrix} = I.$$

Similarly, $\{v_i\}$, $i = 1, ..., r$, are an orthonormal system, so

$$V^\top V = \begin{pmatrix} v_1^\top \\ \vdots \\ v_r^\top \end{pmatrix} \begin{pmatrix} v_1 & \cdots & v_r \end{pmatrix} = \begin{pmatrix} \langle v_1, v_1\rangle & \cdots & \langle v_1, v_r\rangle \\ \vdots & \ddots & \vdots \\ \langle v_r, v_1\rangle & \cdots & \langle v_r, v_r\rangle \end{pmatrix} = I.$$

4.5. From Eq. (3.29), we see that

$$UU^\top = \begin{pmatrix} u_1 & \cdots & u_r \end{pmatrix} \begin{pmatrix} u_1^\top \\ \vdots \\ u_r^\top \end{pmatrix} = \sum_{i=1}^{r} u_i u_i^\top = P_{\mathcal{U}}.$$

Similarly, we see that

$$VV^\top = \begin{pmatrix} u_1 & \cdots & u_r \end{pmatrix} \begin{pmatrix} v_1^\top \\ \vdots \\ v_r^\top \end{pmatrix} = \sum_{i=1}^{r} v_i v_i^\top = P_{\mathcal{V}}.$$

Chapter 5

5.1. The following holds:

$$\begin{aligned} A^- A &= \Big(\sum_{i=1}^{n} \frac{v_i u_i^\top}{\sigma_i}\Big)\Big(\sum_{j=1}^{n} \sigma_j u_j v_j^\top\Big) = \sum_{i,j=1}^{n} \frac{\sigma_i}{\sigma_j} v_i u_i^\top u_j v_j^\top \\ &= \sum_{i,j=1}^{n} \frac{\sigma_i}{\sigma_j} v_i \langle u_i, u_j \rangle v_j^\top \\ &= \sum_{i,j=1}^{n} \frac{\sigma_i}{\sigma_j} \delta_{ij} v_i v_j^\top = \sum_{i=1}^{n} v_i v_j^\top = I. \end{aligned}$$

Since the product is the identity, A^- is the inverse of A.

5.2. We see that

$$\begin{aligned} AA^- &= U \begin{pmatrix} \sigma_1 & & \\ & \ddots & \\ & & \sigma_r \end{pmatrix} V^\top V \begin{pmatrix} 1/\sigma_1 & & \\ & \ddots & \\ & & 1/\sigma_r \end{pmatrix} U^\top \\ &= U \begin{pmatrix} \sigma_1 & & \\ & \ddots & \\ & & \sigma_r \end{pmatrix} \begin{pmatrix} 1/\sigma_1 & & \\ & \ddots & \\ & & 1/\sigma_r \end{pmatrix} U^\top = UU^\top = P_{\mathcal{U}}, \end{aligned}$$

where Eqs. (4.12) and (4.13) are used. Similarly, we see that

$$
\begin{aligned}
A^{-}A &= V\begin{pmatrix} 1/\sigma_1 & & \\ & \ddots & \\ & & 1/\sigma_r \end{pmatrix} U^\top U \begin{pmatrix} \sigma_1 & & \\ & \ddots & \\ & & \sigma_r \end{pmatrix} V^\top \\
&= V\begin{pmatrix} 1/\sigma_1 & & \\ & \ddots & \\ & & 1/\sigma_r \end{pmatrix} \begin{pmatrix} \sigma_1 & & \\ & \ddots & \\ & & \sigma_r \end{pmatrix} V^\top = VV^\top = P_{\mathcal{V}}.
\end{aligned}
$$

5.3. We see that

$$
\begin{aligned}
A^{-}AA^{-} &= V\begin{pmatrix} 1/\sigma_1 & & \\ & \ddots & \\ & & 1/\sigma_r \end{pmatrix} U^\top U \begin{pmatrix} \sigma_1 & & \\ & \ddots & \\ & & \sigma_r \end{pmatrix} \\
&\quad V^\top V \begin{pmatrix} 1/\sigma_1 & & \\ & \ddots & \\ & & 1/\sigma_r \end{pmatrix} U^\top \\
&= V\begin{pmatrix} 1/\sigma_1 & & \\ & \ddots & \\ & & 1/\sigma_r \end{pmatrix} \begin{pmatrix} \sigma_1 & & \\ & \ddots & \\ & & \sigma_r \end{pmatrix} \\
&\quad \begin{pmatrix} 1/\sigma_1 & & \\ & \ddots & \\ & & 1/\sigma_r \end{pmatrix} U^\top \\
&= V\begin{pmatrix} 1/\sigma_1 & & \\ & \ddots & \\ & & 1/\sigma_r \end{pmatrix} U^\top = A^{-},
\end{aligned}
$$

where Eq. (4.12) is used. Similarly, the following holds:

$$
\begin{aligned}
AA^{-}A &= U\begin{pmatrix} \sigma_1 & & \\ & \ddots & \\ & & \sigma_r \end{pmatrix} V^\top V \begin{pmatrix} 1/\sigma_1 & & \\ & \ddots & \\ & & 1/\sigma_r \end{pmatrix} \\
&\quad U^\top U \begin{pmatrix} \sigma_1 & & \\ & \ddots & \\ & & \sigma_r \end{pmatrix} V^\top V \\
&= U\begin{pmatrix} \sigma_1 & & \\ & \ddots & \\ & & \sigma_r \end{pmatrix}\begin{pmatrix} 1/\sigma_1 & & \\ & \ddots & \\ & & 1/\sigma_r \end{pmatrix}\begin{pmatrix} \sigma_1 & & \\ & \ddots & \\ & & \sigma_r \end{pmatrix} V^\top \\
&= U\begin{pmatrix} \sigma_1 & & \\ & \ddots & \\ & & \sigma_r \end{pmatrix} V^\top = A.
\end{aligned}
$$

5.4. For $A = \left(A_{ij}\right)$ and $B = \left(B_{ij}\right)$, the (i, j) elements of AB and BA are $\sum_k A_{ik}B_{kj}$ and $\sum_k B_{ik}A_{kj}$, respectively. Hence, their respective traces are $\sum_{j,k} A_{jk}B_{kj}$ and $\sum_{j,k} B_{jk}A_{kj}$, which are equal.

5.5. We see that

$$
\begin{aligned}
\|AU\|^2 &= \mathrm{tr}(AU(AU)^\top) = \mathrm{tr}(AUU^\top A^\top) = \mathrm{tr}(AA^\top) = \|A\|^2, \\
\|VA\|^2 &= \mathrm{tr}((VA)^\top VA) = \mathrm{tr}(A^\top V VA) = \mathrm{tr}(A^\top A) = \|A\|^2, \\
\|VAU\|^2 &= \mathrm{tr}(VAU(VAU)^\top) = \mathrm{tr}(VAUU^\top A^\top V^\top) = \mathrm{tr}(VAA^\top V^\top) \\
&= \mathrm{tr}(V^\top VAA^\top) = \mathrm{tr}(AA^\top) = \|A\|^2,
\end{aligned}
$$

where we have used Eq. (5.23) and noted that U and V are orthogonal matrices so that $U^\top U = UU^\top = I$ and $V^\top V = VV^\top = I$ hold.

5.6. From Eq. (4.10), we see that

$$\begin{aligned}
\|A\|^2 &= \mathrm{tr}(AA^\top) = \mathrm{tr}(U \begin{pmatrix} \sigma_1 & & \\ & \ddots & \\ & & \sigma_r \end{pmatrix} V^\top V \begin{pmatrix} \sigma_1 & & \\ & \ddots & \\ & & \sigma_r \end{pmatrix} U^\top) \\
&= \mathrm{tr}(U \begin{pmatrix} \sigma_1 & & \\ & \ddots & \\ & & \sigma_r \end{pmatrix} \begin{pmatrix} \sigma_1 & & \\ & \ddots & \\ & & \sigma_r \end{pmatrix} U^\top) \\
&= \mathrm{tr}(U \begin{pmatrix} \sigma_1^2 & & \\ & \ddots & \\ & & \sigma_r^2 \end{pmatrix} U^\top) \\
&= \mathrm{tr}(U^\top U \begin{pmatrix} \sigma_1^2 & & \\ & \ddots & \\ & & \sigma_r^2 \end{pmatrix}) = \mathrm{tr}(\begin{pmatrix} \sigma_1^2 & & \\ & \ddots & \\ & & \sigma_r^2 \end{pmatrix}) = \sigma_1^2 + \cdots + \sigma_r^2,
\end{aligned}$$

where we have noted that U and V are not necessarily orthogonal matrices, i.e., they may not be square matrices, but Eq. (4.12) holds. From this and Eqs. (5.14) and (5.16), we see that

$$A - (A)_r = U \begin{pmatrix} 0 & & & & & \\ & \ddots & & & & \\ & & 0 & & & \\ & & & \sigma_{r+1} & & \\ & & & & \ddots & \\ & & & & & \sigma_l \end{pmatrix} V^\top.$$

Hence, from Eq. (5.24), Eq. (5.20) is obtained.

Chapter 6

6.1. (1) Equation (6.5) is rewritten as

$$\begin{aligned}
J &= \langle Ax - b, Ax - b\rangle \\
&= \langle Ax, Ax\rangle - 2\langle Ax, b\rangle + \langle b, b\rangle \\
&= \langle x, A^\top Ax\rangle - 2\langle x, A^\top b\rangle + \|b\|^2.
\end{aligned}$$

Differentiating this with respect to x and letting the result be (↪ Appendix Eqs. (A.15) and (A.25)), we obtain

$$2A^\top Ax - 2A^\top b = 0.$$

If $m > n$ and $r = n$, the $n \times n$ matrix $\boldsymbol{A}^\top \boldsymbol{A}$ is nonsingular. Hence, the solution $\boldsymbol{x}$ is given by Eq. (6.28).

(2) Replacing $\boldsymbol{A}^\top \boldsymbol{A}\boldsymbol{x}$ in the above expression of J by $\boldsymbol{A}^\top \boldsymbol{b}$, we can write J as

$$J = \langle \boldsymbol{x}, \boldsymbol{A}^\top \boldsymbol{b} \rangle - 2\langle \boldsymbol{x}, \boldsymbol{A}^\top \boldsymbol{b} \rangle + \|\boldsymbol{b}\|^2 = \|\boldsymbol{b}\|^2 - \langle \boldsymbol{x}, \boldsymbol{A}^\top \boldsymbol{b} \rangle.$$

6.2. If $\boldsymbol{A}$ has the singular value decomposition of Eq. (4.4), the following holds:

$$\begin{aligned} \boldsymbol{A}^\top \boldsymbol{A} &= \Big(\sum_{i=1}^{r} \sigma_i \boldsymbol{v}_i \boldsymbol{u}_i^\top\Big)\Big(\sum_{j=1}^{r} \sigma_j \boldsymbol{u}_j \boldsymbol{v}_j^\top\Big) = \sum_{i,j=1}^{r} \sigma_i \sigma_j \boldsymbol{v}_i \boldsymbol{u}_i^\top \boldsymbol{u}_j \boldsymbol{v}_j^\top \\ &= \sum_{i,j=1}^{r} \sigma_i \sigma_j \boldsymbol{v}_i \langle \boldsymbol{u}_i, \boldsymbol{u}_j \rangle \boldsymbol{v}_j^\top = \sum_{i,j=1}^{r} \delta_{ij} \sigma_i \sigma_j \boldsymbol{v}_i \boldsymbol{v}_j^\top = \sum_{i=1}^{r} \sigma_i^2 \boldsymbol{v}_i \boldsymbol{v}_i^\top, \end{aligned}$$

$$\begin{aligned} (\boldsymbol{A}^\top \boldsymbol{A})^{-1} \boldsymbol{A}^\top &= \Big(\sum_{i=1}^{r} \frac{\boldsymbol{v}_i \boldsymbol{v}_i^\top}{\sigma_i^2}\Big)\Big(\sum_{j=1}^{r} \sigma_j \boldsymbol{v}_j \boldsymbol{u}_j^\top\Big) = \sum_{i,j=1}^{r} \frac{\sigma_j}{\sigma_i^2} \boldsymbol{v}_i \boldsymbol{v}_i^\top \boldsymbol{v}_j \boldsymbol{u}_j^\top \\ &= \sum_{i,j=1}^{r} \frac{\sigma_j}{\sigma_i^2} \boldsymbol{v}_i \langle \boldsymbol{v}_i, \boldsymbol{v}_j \rangle \boldsymbol{u}_j^\top = \sum_{i,j=1}^{r} \frac{\sigma_j}{\sigma_i^2} \delta_{ij} \boldsymbol{v}_i \boldsymbol{u}_j^\top = \sum_{i=1}^{r} \frac{\boldsymbol{v}_i \boldsymbol{u}_i^\top}{\sigma_i} = \boldsymbol{A}^-, \end{aligned}$$

6.3. We minimize $\|\boldsymbol{x}\|^2/2$ subject to $\boldsymbol{A}\boldsymbol{x} = \boldsymbol{b}$, where the coefficient 1/2 is only formal. Introducing the Lagrange multiplier $\boldsymbol{\lambda}$ ($\hookrightarrow$ Appendix Eq. (A.48)), consider

$$\frac{1}{2}\|\boldsymbol{x}\|^2 - \langle \boldsymbol{\lambda}, \boldsymbol{A}\boldsymbol{x} - \boldsymbol{b} \rangle = \frac{1}{2}\langle \boldsymbol{x}, \boldsymbol{x} \rangle - \langle \boldsymbol{A}^\top \boldsymbol{\lambda}, \boldsymbol{x} \rangle + \langle \boldsymbol{\lambda}, \boldsymbol{b} \rangle.$$

Differentiating this with respect to $\boldsymbol{x}$ ($\hookrightarrow$ Appendix Eqs. (A.15) and (A.25)) and letting the result be $\boldsymbol{0}$, we obtain

$$\boldsymbol{x} - \boldsymbol{A}^\top \boldsymbol{\lambda} = \boldsymbol{0}.$$

Hence, the equation $\boldsymbol{A}\boldsymbol{x} = \boldsymbol{b}$ is written as

$$\boldsymbol{A}\boldsymbol{A}^\top \boldsymbol{\lambda} = \boldsymbol{b}.$$

If $m < n$ and $r = m$, the $m \times m$ matrix $\boldsymbol{A}\boldsymbol{A}^\top$ is nonsingular, so $\boldsymbol{\lambda}$ is given by

$$\boldsymbol{\lambda} = (\boldsymbol{A}\boldsymbol{A}^\top)^{-1} \boldsymbol{b}.$$

Hence, $\boldsymbol{x}$ is given in the form

$$\boldsymbol{x} = \boldsymbol{A}^\top \boldsymbol{\lambda} = \boldsymbol{A}^\top (\boldsymbol{A}\boldsymbol{A}^\top)^{-1} \boldsymbol{b}.$$

Since $\boldsymbol{A}\boldsymbol{x} = \boldsymbol{b}$ is satisfied, the residual is $J = \|\boldsymbol{A}\boldsymbol{x} - \boldsymbol{b}\|^2 = 0$.

6.4. If $\boldsymbol{A}$ has the singular value decomposition of Eq. (4.4), the following holds:

$$
\begin{aligned}
\boldsymbol{A}\boldsymbol{A}^\top &= \Big(\sum_{i=1}^r \sigma_i \boldsymbol{u}_i \boldsymbol{v}_i^\top\Big)\Big(\sum_{j=1}^r \sigma_j \boldsymbol{v}_j \boldsymbol{u}_j^\top\Big) = \sum_{i,j=1}^r \sigma_i\sigma_j \boldsymbol{u}_i \boldsymbol{v}_i^\top \boldsymbol{v}_j \boldsymbol{u}_j^\top \\
&= \sum_{i,j=1}^r \sigma_i\sigma_j \boldsymbol{u}_i \langle \boldsymbol{v}_i, \boldsymbol{v}_j\rangle \boldsymbol{u}_j^\top = \sum_{i,j=1}^r \delta_{ij}\sigma_i\sigma_j \boldsymbol{u}_i \boldsymbol{u}_j^\top = \sum_{i=1}^r \sigma_i^2 \boldsymbol{u}_i \boldsymbol{u}_i^\top,
\end{aligned}
$$

$$
\begin{aligned}
\boldsymbol{A}^\top(\boldsymbol{A}\boldsymbol{A}^\top)^{-1} &= \Big(\sum_{i=1}^r \sigma_i \boldsymbol{v}_i \boldsymbol{u}_i^\top\Big)\Big(\sum_{j=1}^r \frac{\boldsymbol{u}_j \boldsymbol{u}_j^\top}{\sigma_j^2}\Big) = \sum_{i,j=1}^r \frac{\sigma_i}{\sigma_j^2} \boldsymbol{v}_i \boldsymbol{u}_i^\top \boldsymbol{u}_j \boldsymbol{u}_j^\top \\
&= \sum_{i,j=1}^r \frac{\sigma_i}{\sigma_j^2} \boldsymbol{v}_i \langle \boldsymbol{u}_i, \boldsymbol{u}_j\rangle \boldsymbol{u}_j^\top = \sum_{i,j=1}^r \frac{\sigma_i}{\sigma_j^2}\delta_{ij} \boldsymbol{v}_i \boldsymbol{u}_j^\top = \sum_{i=1}^r \frac{\boldsymbol{v}_i \boldsymbol{u}_i^\top}{\sigma_i} = \boldsymbol{A}^-.
\end{aligned}
$$

6.5. Differentiating Eq. (6.19) with respect to x, we obtain

$$
\begin{aligned}
\frac{dJ}{dx} &= 2(a_1x - b_1)a_1 + \cdots + 2(a_mx - b_m)a_m \\
&= 2(a_1^1 + \cdots + a_m^2)x - 2(a_1b_1 + \cdots + a_mb_m).
\end{aligned}
$$

Letting this be 0, we obtain Eq. (6.17).

6.6. We minimize $\|\boldsymbol{x}\|^2/2$ subject to Eq. (6.20), where the coefficient 1/2 is only formal. Introducing the Lagrange multiplier λ ($\hookrightarrow$ Appendix Eq. (A.46)), consider

$$
\frac{1}{2}\|\boldsymbol{x}\|^2 - \lambda(\langle \boldsymbol{a}, \boldsymbol{x}\rangle - b).
$$

Differentiating this with respect to $\boldsymbol{x}$ ($\hookrightarrow$ Appendix Eqs. (A.15) and, (A.25)) and letting the result be $\boldsymbol{0}$, we obtain

$$
\boldsymbol{x} - \lambda\boldsymbol{a} = \boldsymbol{0},
$$

i.e., $\boldsymbol{x} = \lambda\boldsymbol{a}$. Substituting this into $\langle \boldsymbol{a}, \boldsymbol{x}\rangle = b$, we obtain

$$
\lambda\|\boldsymbol{a}\|^2 = b,
$$

i.e., $\lambda = b/\|\boldsymbol{a}\|^2$. Hence, we obtain

$$
\boldsymbol{x} = \frac{b\boldsymbol{a}}{\|\boldsymbol{a}\|^2}.
$$

Chapter 7

7.1. The (i, i) element Σ_{ii} is, by definition, $E[\Delta x_i^2] = E[(x_i - \bar{x}_i)^2]$, which gives the variance of x_i. The non-diagonal element $\Sigma_{ij}, i \neq j$ is $E[\Delta x_i \Delta x_j] = E[(x_i - \bar{x}_i)(x_j - \bar{x}_j)]$, which gives the covariance of x_i and x_j. Note that in this chapter we are assuming that the expectation of x_i is $\bar{x}_i = 0$.

7.2. Evidently, $\boldsymbol{X}$ is a symmetric matrix: $\boldsymbol{X}^\top = \boldsymbol{x}\boldsymbol{x}^\top = (\boldsymbol{x}^\top)^\top \boldsymbol{x}^\top = \boldsymbol{X}$. Let σ be its eigenvalue, and $\boldsymbol{u}$ the corresponding eigenvector. Computing the inner product of $\boldsymbol{X}\boldsymbol{u} = \sigma\boldsymbol{u}$ with $\boldsymbol{u}$ on both sides, we obtain

$$\langle \boldsymbol{u}, \boldsymbol{X}\boldsymbol{u} \rangle = \sigma \langle \boldsymbol{u}, \boldsymbol{u} \rangle = \sigma \|\boldsymbol{u}\|^2,$$

but since

$$\langle \boldsymbol{u}, \boldsymbol{X}\boldsymbol{u} \rangle = \langle \boldsymbol{u}, \boldsymbol{x}\boldsymbol{x}^\top \boldsymbol{u} \rangle = \langle \boldsymbol{u}, \boldsymbol{x} \rangle \langle \boldsymbol{x}, \boldsymbol{u} \rangle = \langle \boldsymbol{u}, \boldsymbol{x} \rangle^2 \geq 0,$$

we see that $\sigma \geq 0$. Similarly, for multiple vectors we see that

$$\begin{aligned} \langle \boldsymbol{u}, \boldsymbol{X}\boldsymbol{u} \rangle &= \langle \boldsymbol{u}, \sum_{\alpha=1}^{N} \boldsymbol{x}_\alpha \boldsymbol{x}_\alpha^\top \boldsymbol{u} \rangle = \sum_{\alpha=1}^{N} \langle \boldsymbol{u}, \boldsymbol{x}_\alpha \boldsymbol{x}_\alpha^\top \boldsymbol{u} \rangle = \sum_{\alpha=1}^{N} \langle \boldsymbol{u}, \boldsymbol{x}_\alpha \rangle \langle \boldsymbol{x}_\alpha, \boldsymbol{u} \rangle \\ &= \sum_{\alpha=1}^{N} \langle \boldsymbol{u}, \boldsymbol{x}_\alpha \rangle^2 \geq 0, \end{aligned}$$

and hence $\sigma \geq 0$.

7.3. From Eq. (2.28), we see that $\mathrm{tr}(\boldsymbol{x}\boldsymbol{x}^\top) = \langle \boldsymbol{x}, \boldsymbol{x} \rangle = \|\boldsymbol{x}\|^2$. Hence, $\mathrm{tr}(\sum_{\alpha=1}^{N} \boldsymbol{x}_\alpha \boldsymbol{x}_\alpha^\top) = \sum_{\alpha=1}^{N} \|\boldsymbol{x}_\alpha\|^2$.

7.4. Let $\boldsymbol{\Sigma} = \mathrm{diag}(\sigma_1^2, \sigma_2^2, \sigma_3^2)$, $\sigma_1^2, \sigma_2^2, \sigma_3^2 > 0$, in 3D, where $\mathrm{diag}(\cdots)$ denotes the diagonal matrix with diagonal elements $\cdots$ in that order. Its inverse is $\boldsymbol{\Sigma}^{-1} = \mathrm{diag}(1/\sigma_1^2, 1/\sigma_2^2, 1/\sigma_3^2)$, so that Eq. (7.10) is written as

$$\frac{(x - \bar{x})^2}{\sigma_1^2} + \frac{(y - \bar{y})^2}{\sigma_2^2} + \frac{(z - \bar{z})^2}{\sigma_3^2} = 1.$$

This describes an ellipsoid centered on $(\bar{x}, \bar{y}, \bar{z})$ with the coordinate axes as the axes of symmetry, having radii σ_1, σ_2, and σ_3 along the x-, y-, and z-axes, respectively.

7.5. The covariance matrix $\boldsymbol{\Sigma}$, which is assumed positive definite here, has the spectral decomposition $\boldsymbol{\Sigma} = \sum_{i=1}^{n} \sigma_i^2 \boldsymbol{u}_i \boldsymbol{u}_i^\top$, $\sigma_i^2 > 0$, $i = 1, ..., n$. Translate the coordinate system so that $\bar{\boldsymbol{x}}$ coincides with the origin O and rotate it so that the coordinate axes coincide with the eigenvectors $\boldsymbol{u}_1, ..., \boldsymbol{u}_n$ of $\boldsymbol{\Sigma}$. With respect to this new coordinate system, the covariance matrix has the form $\boldsymbol{\Sigma} = \mathrm{diag}(\sigma_1^2, ..., \sigma_n^2)$, and its inverse is $\boldsymbol{\Sigma}^{-1} = \mathrm{diag}(1/\sigma_1^2, ..., 1/\sigma_n^2)$. Hence, Eq. (7.10) is now written as

$$\frac{x_1^2}{\sigma_1^2} + \cdots + \frac{x_n^2}{\sigma_n^2} = 1.$$

This describes an ellipsoid centered on the origin with the coordinate axes as the axes of symmetry, having the radius σ_i in each coordinate axis direction. This means that, with respect to the original coordinate system, the ellipsoid is centered on the expectation $\bar{\boldsymbol{x}}$ with the eigenvectors $\boldsymbol{u}_i$ of $\boldsymbol{\Sigma}$ as the axes of symmetry, having the radius σ_i in each axis direction.

7.6. The (i, i) element of $\boldsymbol{S}$

$$S_{ii} = \frac{1}{N}\sum_{\alpha=1}^{N}(\hat{x}_{i\alpha} - m_i)^2$$

is, by definition, the sample variance of $x_{i\alpha}$, where

$$m_i = \frac{1}{N}\sum_{\alpha=1}^{N}\hat{x}_{i\alpha}$$

is the sample mean of $x_{i\alpha}$. The non-diagonal element

$$S_{ij} = \frac{1}{N}\sum_{\alpha=1}^{N}(\hat{x}_{i\alpha} - m_i)(\hat{x}_{j\alpha} - m_j)$$

is the covariance of $x_{i\alpha}$ and $x_{j\alpha}$.

Chapter 8

8.1. Let $\boldsymbol{A} = \sum_{i=1}^{n}\lambda_i \boldsymbol{u}_i \boldsymbol{u}_i^\top$ be the spectral decomposition of $\boldsymbol{A}$. Then,

$$\mathrm{tr}\boldsymbol{A} = \sum_{i=1}^{n}\lambda_i \mathrm{tr}(\boldsymbol{u}_i \boldsymbol{u}_i^\top) = \sum_{i=1}^{n}\lambda_i \|\boldsymbol{u}_i\|^2 = \sum_{i=1}^{n}\lambda_i.$$

Recall that the eigenvectors $\boldsymbol{u}_i$ are all unit vectors.

8.2. If we take $\boldsymbol{x}_0$ as a reference and regard it as the origin, the remaining n vectors are linearly independent if and only if

$$\begin{vmatrix} \boldsymbol{x}_1 - \boldsymbol{x}_0 & \cdots & \boldsymbol{x}_n - \boldsymbol{x}_0 \end{vmatrix} \neq 0.$$

The $n \times n$ determinant on the left side is written as the $(n+1) \times (n+1)$ determinant

$$\begin{vmatrix} x_0 & x_1 - x_0 & \cdots & x_n - x_0 \\ 1 & 0 & \cdots & 0 \end{vmatrix} = \begin{vmatrix} x_0 & x_1 & \cdots & x_n \\ 1 & 1 & \cdots & 1 \end{vmatrix},$$

where we have added the first row to the other rows, which does not change the determinant. Here, we chose $\boldsymbol{x}_0$ as the reference, but we obtain the same result whichever $\boldsymbol{x}_i$ we choose.

8.3. Differentiating $J = \sum_{\alpha=1}^{N} \|\boldsymbol{x}_\alpha - \boldsymbol{g}\|^2 = \sum_{\alpha=1}^{N}(\boldsymbol{x}_\alpha - \boldsymbol{g}, \boldsymbol{x}_\alpha - \boldsymbol{g})$ with respect to $\boldsymbol{x}$, we obtain $\nabla_{\boldsymbol{x}} J = 2\sum_{\alpha=1}^{N}(\boldsymbol{x}_\alpha - \boldsymbol{g}) = 2\sum_{\alpha=1}^{N} \boldsymbol{x}_\alpha - 2N\boldsymbol{g}$ ($\hookrightarrow$ Appendix Eqs. (A.15) and (A.25)). Letting this be $\nabla_{\boldsymbol{x}} J = 2\sum_{\alpha=1}^{N}(\boldsymbol{x}_\alpha - \boldsymbol{g}) = \boldsymbol{0}$, we obtain Eq. (8.13).

8.4. From $\sum_{\alpha=1}^{N} \boldsymbol{x}_\alpha = N\boldsymbol{g}$, we see that

$$\begin{aligned} \boldsymbol{\Sigma} &= \sum_{\alpha=1}^{N}(\boldsymbol{x}_\alpha - \boldsymbol{g})(\boldsymbol{x}_\alpha - \boldsymbol{g})^\top = \sum_{\alpha=1}^{N} \boldsymbol{x}_\alpha \boldsymbol{x}_\alpha^\top - \sum_{\alpha=1}^{N} \boldsymbol{x}_\alpha \boldsymbol{g}^\top - \sum_{\alpha=1}^{N} \boldsymbol{g}\boldsymbol{x}_\alpha^\top + \sum_{\alpha=1}^{N} \boldsymbol{g}\boldsymbol{g}^\top \\ &= \sum_{\alpha=1}^{N} \boldsymbol{x}_\alpha \boldsymbol{x}_\alpha^\top - N\boldsymbol{g}\boldsymbol{g}^\top - N\boldsymbol{g}\boldsymbol{g}^\top + N\boldsymbol{g}\boldsymbol{g}^\top = \sum_{\alpha=1}^{N} \boldsymbol{x}_\alpha \boldsymbol{x}_\alpha^\top - N\boldsymbol{g}\boldsymbol{g}^\top. \end{aligned}$$

Chapter 9

9.1. If $\boldsymbol{A}$ has rank r or less, we can factorize it in the form $\boldsymbol{A} = \boldsymbol{A}_1\boldsymbol{A}_2$ for some $m \times r$ matrix $\boldsymbol{A}_1$ and some $r \times n$ matrix $\boldsymbol{A}_2$ as shown by Eqs. (9.4)–(9.8). Conversely, if we can write $\boldsymbol{A} = \boldsymbol{A}_1\boldsymbol{A}_2$ for some $m \times r$ matrix $\boldsymbol{A}_1$ and some $r \times n$ matrix $\boldsymbol{A}_2$, Eq. (9.3) implies that $\boldsymbol{A}$ has r or less.

9.2. (1) If we write the three columns of the $2M$ motion matrix $\boldsymbol{M}$ of Eq. (9.13) as $\boldsymbol{m}_1$, $\boldsymbol{m}_2$, and $\boldsymbol{m}_3$, Eq. (9.14) implies that the αth column of Eq. (9.12) is written as

$$\begin{pmatrix} x_{\alpha 1} \\ y_{\alpha 1} \\ \cdots \\ x_{\alpha M} \\ y_{\alpha M} \end{pmatrix} = X_\alpha \boldsymbol{m}_1 + Y_\alpha \boldsymbol{m}_2 + Z_\alpha \boldsymbol{m}_3.$$

This means that the trajectory of the αth point is included in the 3D subspace spanned by $\boldsymbol{m}_1$, $\boldsymbol{m}_2$, and $\boldsymbol{m}_3$. Hence, the trajectory of any point is included in this subspace.

(2) Computation of this 3D subspace reduces to the problem of fitting a 3D subspace to N points $(x_{\alpha 1}, y_{\alpha 1}, \ldots, x_{\alpha M}, y_{\alpha M})$, $\alpha = 1, \ldots, N$, in $2M$ dimensions. This can be

done, as described in Sec. 8.3, by computing the singular value decomposition of the $2M \times N$ matrix having columns that represent the N points. This matrix is nothing but the observation matrix $\boldsymbol{W}$ of Eq. (9.12). Hence, if we compute its singular value decomposition in the form

$$\boldsymbol{W} = \sigma_1 \boldsymbol{u}_1 \boldsymbol{v}_1 + \sigma_2 \boldsymbol{u}_2 \boldsymbol{v}_2 + \sigma_3 \boldsymbol{u}_3 \boldsymbol{v}_3 + \cdots ,$$

the three vectors $\{\boldsymbol{u}_1, \boldsymbol{u}_2, \boldsymbol{u}_3\}$ give an orthonormal basis of that 3D subspace. Note that Eq. (9.14) implies that the matrix $\boldsymbol{W}$ has rank 3 and hence $\sigma_4 = \sigma_5 = \cdots = 0$, but this holds only for hypothetical affine cameras. For the observation matrix $\boldsymbol{W}$ obtained using real cameras, the singular values $\sigma_4, \sigma_5, \cdots$ are not necessarily 0, so we truncate these and use the first three terms, which gives an optimal fitting.

Chapter 10

10.1. Consider, for example, the determinant consisting of the first, second, third, and fifth rows of Eq. (10.5):

$$\begin{vmatrix} f_0 P_{0(11)} - x_0 P_{0(31)} & f_0 P_{0(12)} - x_0 P_{0(32)} & f_0 P_{0(13)} - x_0 P_{0(33)} & f_0 P_{0(14)} - x_0 P_{0(34)} \\ f_0 P_{0(21)} - y_0 P_{0(31)} & f_0 P_{0(22)} - y_0 P_{0(32)} & f_0 P_{0(23)} - y_0 P_{0(33)} & f_0 P_{0(24)} - y_0 P_{0(34)} \\ f_0 P_{1(11)} - x_1 P_{1(31)} & f_0 P_{1(12)} - x_1 P_{1(32)} & f_0 P_{1(13)} - x_1 P_{1(33)} & f_0 P_{1(14)} - x_1 P_{1(34)} \\ f_0 P_{2(11)} - x_2 P_{2(31)} & f_0 P_{2(12)} - x_2 P_{2(32)} & f_0 P_{2(13)} - x_2 P_{2(33)} & f_0 P_{2(14)} - x_2 P_{2(34)} \end{vmatrix}$$
$$= 0.$$

Dividing each element by f_0 and writing x_0/f_0, y_0/f_0, x_1/f_0, and x_2/f_0 as $x_{0(1)}$, $x_{0(2)}$, $x_{1(1)}$, and $x_{2(1)}$, respectively, we can rewrite the above equation as

$$\begin{vmatrix} P_{0(11)} - x_{0(1)} P_{0(31)} & P_{0(12)} - x_{0(1)} P_{0(32)} & P_{0(13)} - x_{0(1)} P_{0(33)} & P_{0(14)} - x_{0(1)} P_{0(34)} \\ P_{0(21)} - x_{0(2)} P_{0(31)} & P_{0(22)} - x_{0(2)} P_{0(32)} & P_{0(23)} - x_{0(2)} P_{0(33)} & P_{0(24)} - x_{0(2)} P_{0(34)} \\ P_{1(11)} - x_{1(1)} P_{1(31)} & P_{1(12)} - x_{1(1)} P_{1(32)} & P_{1(13)} - x_{1(1)} P_{1(33)} & P_{1(14)} - x_{1(1)} P_{1(34)} \\ P_{2(11)} - x_{2(1)} P_{2(31)} & P_{2(12)} - x_{2(1)} P_{2(32)} & P_{2(13)} - x_{2(1)} P_{2(33)} & P_{2(14)} - x_{2(1)} P_{2(34)} \end{vmatrix}$$

$$= \begin{vmatrix} P_{0(11)} - x_{0(1)} P_{0(31)} & P_{0(12)} - x_{0(1)} P_{0(32)} & P_{0(13)} - x_{0(1)} P_{0(33)} \\ P_{0(21)} - x_{0(2)} P_{0(31)} & P_{0(22)} - x_{0(2)} P_{0(32)} & P_{0(23)} - x_{0(2)} P_{0(33)} \\ P_{1(11)} - x_{1(1)} P_{1(31)} & P_{1(12)} - x_{1(1)} P_{1(32)} & P_{1(13)} - x_{1(1)} P_{1(33)} \\ P_{2(11)} - x_{2(1)} P_{2(31)} & P_{2(12)} - x_{2(1)} P_{2(32)} & P_{2(13)} - x_{2(1)} P_{2(33)} \\ P_{0(31)} & P_{0(32)} & P_{0(33)} \\ P_{1(31)} & P_{1(32)} & P_{1(33)} \\ P_{2(31)} & P_{2(32)} & P_{2(33)} \end{vmatrix}$$

$$
\left.\begin{matrix}
P_{0(14)}-x_{0(1)}P_{0(34)} & 0 & 0 & 0 \\
P_{0(24)}-x_{0(2)}P_{0(34)} & 0 & 0 & 0 \\
P_{1(14)}-x_{1(1)}P_{1(34)} & 0 & 0 & 0 \\
P_{2(14)}-x_{2(1)}P_{2(34)} & 0 & 0 & 0 \\
P_{0(34)} & 1 & 0 & 0 \\
P_{1(34)} & 0 & 1 & 0 \\
P_{2(34)} & 0 & 0 & 1
\end{matrix}\right|
$$

$$
= \begin{vmatrix}
P_{0(11)} & P_{0(12)} & P_{0(13)} & P_{0(14)} & x_{0(1)} & 0 & 0 \\
P_{0(21)} & P_{0(22)} & P_{0(23)} & P_{0(24)} & x_{0(2)} & 0 & 0 \\
P_{1(11)} & P_{1(12)} & P_{1(13)} & P_{1(14)} & 0 & x_{1(1)} & 0 \\
P_{2(11)} & P_{2(12)} & P_{2(13)} & P_{2(14)} & 0 & 0 & x_{2(1)} \\
P_{0(31)} & P_{0(32)} & P_{0(33)} & P_{0(34)} & 1 & 0 & 0 \\
P_{1(31)} & P_{1(32)} & P_{1(33)} & P_{1(34)} & 0 & 1 & 0 \\
P_{2(31)} & P_{2(32)} & P_{2(33)} & P_{2(34)} & 0 & 0 & 1
\end{vmatrix}
$$

$$
= \begin{vmatrix}
P_{0(21)} & P_{0(22)} & P_{0(23)} & P_{0(24)} & 0 & 0 \\
P_{1(11)} & P_{1(12)} & P_{1(13)} & P_{1(14)} & x_{1(1)} & 0 \\
P_{2(11)} & P_{2(12)} & P_{2(13)} & P_{2(14)} & 0 & x_{2(1)} \\
P_{0(31)} & P_{0(32)} & P_{0(33)} & P_{0(34)} & 0 & 0 \\
P_{1(31)} & P_{1(32)} & P_{1(33)} & P_{1(34)} & 1 & 0 \\
P_{2(31)} & P_{2(32)} & P_{2(33)} & P_{2(34)} & 0 & 1
\end{vmatrix} x_{0(1)}
$$

$$
- \begin{vmatrix}
P_{0(11)} & P_{0(12)} & P_{0(13)} & P_{0(14)} & 0 & 0 \\
P_{1(11)} & P_{1(12)} & P_{1(13)} & P_{1(14)} & x_{1(1)} & 0 \\
P_{2(11)} & P_{2(12)} & P_{2(13)} & P_{2(14)} & 0 & x_{2(1)} \\
P_{0(31)} & P_{0(32)} & P_{0(33)} & P_{0(34)} & 0 & 0 \\
P_{1(31)} & P_{1(32)} & P_{1(33)} & P_{1(34)} & 1 & 0 \\
P_{2(31)} & P_{2(32)} & P_{2(33)} & P_{2(34)} & 0 & 1
\end{vmatrix} x_{0(2)}
$$

$$
+ \begin{vmatrix}
P_{0(11)} & P_{0(12)} & P_{0(13)} & P_{0(14)} & 0 & 0 \\
P_{0(21)} & P_{0(22)} & P_{0(23)} & P_{0(24)} & 0 & 0 \\
P_{1(11)} & P_{1(12)} & P_{1(13)} & P_{1(14)} & x_{1(1)} & 0 \\
P_{2(11)} & P_{2(12)} & P_{2(13)} & P_{2(14)} & 0 & x_{2(1)} \\
P_{1(31)} & P_{1(32)} & P_{1(33)} & P_{1(34)} & 1 & 0 \\
P_{2(31)} & P_{2(32)} & P_{2(33)} & P_{2(34)} & 0 & 1
\end{vmatrix} = 0, \qquad (*)
$$

where in going from the first expression to the second we have used the fact that the determinant does not change by adding a diagonal block consisting of the identity matrix if either of the corresponding non-diagonal blocks consists of only 0. In going from the second to third expression, the fifth row is multiplied by $x_{0(1)}$ and added to the first row, the sixth row is multiplied by $x_{0(2)}$ and added to the second row, and the seventh row is multiplied by $x_{2(1)}$ and added to the fourth row. The fourth expression is obtained by cofactor expansion of the third expression with respect to the fifth column. The first term of the fourth expression is rewritten by cofactor expansion with respect to the fifth and sixth columns in the following form:

$$
\left(-\begin{vmatrix} p_{0(21)} & p_{0(22)} & p_{0(23)} & p_{0(24)} & 0 \\ p_{2(11)} & p_{2(12)} & p_{2(13)} & p_{2(14)} & x_{2(1)} \\ p_{0(31)} & p_{0(32)} & p_{0(33)} & p_{0(34)} & 0 \\ p_{1(31)} & p_{1(32)} & p_{1(33)} & p_{1(34)} & 0 \\ p_{2(31)} & p_{2(32)} & p_{2(33)} & p_{2(34)} & 1 \end{vmatrix} x_{1(1)}\right.
$$

$$
\left.+\begin{vmatrix} p_{0(21)} & p_{0(22)} & p_{0(23)} & p_{0(24)} & 0 \\ p_{1(11)} & p_{1(12)} & p_{1(13)} & p_{1(14)} & 0 \\ p_{2(11)} & p_{2(12)} & p_{2(13)} & p_{2(14)} & x_{2(1)} \\ p_{0(31)} & p_{0(32)} & p_{0(33)} & p_{0(34)} & 0 \\ p_{2(31)} & p_{2(32)} & p_{2(33)} & p_{2(34)} & 1 \end{vmatrix}\right) x_{0(1)}
$$

$$
= \left(\begin{vmatrix} p_{0(21)} & p_{0(22)} & p_{0(23)} & p_{0(24)} \\ p_{0(31)} & p_{0(32)} & p_{0(33)} & p_{0(34)} \\ p_{1(31)} & p_{1(32)} & p_{1(33)} & p_{1(34)} \\ p_{2(31)} & p_{2(32)} & p_{2(33)} & p_{2(34)} \end{vmatrix} x_{2(1)}\right.
$$

$$
\left.-\begin{vmatrix} p_{0(21)} & p_{0(22)} & p_{0(23)} & p_{0(24)} \\ p_{2(11)} & p_{2(12)} & p_{2(13)} & p_{2(14)} \\ p_{0(31)} & p_{0(32)} & p_{0(33)} & p_{0(34)} \\ p_{1(31)} & p_{1(32)} & p_{1(33)} & p_{1(34)} \end{vmatrix}\right) x_{0(1)} x_{1(1)}
$$

$$+\left(\begin{vmatrix} P_{0(21)} & P_{0(22)} & P_{0(23)} & P_{0(24)} \\ P_{1(11)} & P_{1(12)} & P_{1(13)} & P_{1(14)} \\ P_{0(31)} & P_{0(32)} & P_{0(33)} & P_{0(34)} \\ P_{2(31)} & P_{2(32)} & P_{2(33)} & P_{2(34)} \end{vmatrix} x_{2(1)}\right.$$

$$\left. + \begin{vmatrix} P_{0(21)} & P_{0(22)} & P_{0(23)} & P_{0(24)} \\ P_{1(11)} & P_{1(12)} & P_{1(13)} & P_{1(14)} \\ P_{2(11)} & P_{2(12)} & P_{2(13)} & P_{2(14)} \\ P_{0(31)} & P_{0(32)} & P_{0(33)} & P_{0(34)} \end{vmatrix} \right) x_{0(1)}$$

$$= T_1^{33} x_{0(1)} x_{1(1)} x_{2(1)} - T_1^{31} x_{0(1)} x_{1(1)} - T_1^{13} x_{0(1)} x_{2(1)} + T_1^{11} x_{0(1)}$$

$$= T_1^{33} x_{0(1)} x_{1(1)} x_{2(1)} - T_1^{31} x_{0(1)} x_{1(1)} x_{2(3)} - T_1^{13} x_{0(1)} x_{1(3)} x_{2(1)} + T_1^{11} x_{0(1)} x_{1(3)} x_{2(3)}$$

$$= \sum_{j,k,l,m=1}^{3} \epsilon_{lj2} \epsilon_{mk2} T_1^{lm} x_{0(1)} x_{1(j)} x_{2(k)}.$$

Here, we have used the definition of T_i^{jk} of Eq. (10.7), the fact that $x_{\kappa(3)} = 1$, and the properties of the permutation signature ϵ_{ijk}. Similarly expanding the second and third terms of Eq. $(*)$, we obtain, respectively, expressions

$$\sum_{j,k,l,m=1}^{3} \epsilon_{lj2} \epsilon_{mk2} T_2^{lm} x_{0(1)} x_{1(j)} x_{2(k)}, \qquad \sum_{j,k,l,m=1}^{3} \epsilon_{lj2} \epsilon_{mk2} T_3^{lm} x_{0(1)} x_{1(j)} x_{2(k)}.$$

Thus, Eq. $(*)$ represents the trilinear constraint of Eq. (10.6) for $p = 2$ and $q = 2$, i.e.,

$$\sum_{i,j,k,l,m=1}^{3} \epsilon_{lj2} \epsilon_{mk2} T_i^{lm} x_{0(1)} x_{1(j)} x_{2(k)} = 0.$$

This is the result of extracting the first, second, third, and fifth rows of Eq. (10.5). Extracting other combinations of the rows, we obtain the trilinear constraint for other p and q.

10.2. If we let

$$T_{pq} = \sum_{i,j,k,l,m=1}^{3} \epsilon_{ljp} \epsilon_{mkq} T_i^{lm} x_{0(i)} x_{1(j)} x_{2(k)},$$

the following summation vanishes:

$$\sum_{p=1}^{3} T_{pq} x_{1(p)} = \sum_{i,j,k,l,m,p=1}^{3} \epsilon_{ljp} \epsilon_{mkq} T_i^{lm} x_{0(i)} x_{1(j)} x_{2(k)} x_{1(p)} = 0.$$

This is because the sign of ϵ_{ljp} is reversed by interchanging j and p but the value of $x_{1(j)}x_{1(p)}$ remains the same. Hence, only two among T_{1q}, T_{2q}, and T_{3q} are linearly independent (e.g., T_{3q} can be expressed as a linear combination of T_{1q} and T_{2q}). Similarly, the following holds:

$$\sum_{q=1}^{3} T_{pq}x_{2(q)} = \sum_{i,j,k,l,m,p=1}^{3} \epsilon_{ljp}\epsilon_{mkq}T_i^{lm}x_{0(i)}x_{1(j)}x_{2(k)}x_{2(q)} = 0.$$

Hence, only two among T_{p1}, T_{p2}, and T_{p3} are linearly independent. Thus, only four among nine T_{pq} are linearly independent (e.g., each of T_{13}, T_{23}, T_{31}, T_{32}, and T_{33} can be expressed as a linear combination of T_{11}, T_{12}, T_{21}, and T_{22}).

10.3. Suppose each observed position $\boldsymbol{x}_\kappa$ is already corrected to $\hat{\boldsymbol{x}}_\kappa$ (initially, we let $\hat{\boldsymbol{x}}_\kappa = \boldsymbol{x}_\kappa$). We consider how to correct it to an optimal position $\bar{\boldsymbol{x}}_\kappa$. Instead of directly estimating $\bar{\boldsymbol{x}}_\kappa$, we let

$$\bar{\boldsymbol{x}}_\kappa = \hat{\boldsymbol{x}}_\kappa - \Delta\hat{\boldsymbol{x}}_\kappa, \qquad \kappa = 0, 1, 2,$$

and estimate the additional correction $\Delta\hat{\boldsymbol{x}}_\kappa$. The reprojection error E of Eq. (10.10) is written as

$$E = \sum_{\kappa=0}^{2} \|\tilde{\boldsymbol{x}}_\kappa + \Delta\hat{\boldsymbol{x}}_\kappa\|^2, \qquad (*1)$$

where

$$\tilde{\boldsymbol{x}}_\kappa = \boldsymbol{x}_\kappa - \hat{\boldsymbol{x}}_\kappa, \qquad \kappa = 0, 1, 2.$$

The trilinear constraint of Eq. (10.6) is written as

$$\sum_{i,j,k,l,m=1}^{3} \epsilon_{ljp}\epsilon_{mkq}T_i^{lm}(\hat{x}_{0(i)} - \Delta\hat{x}_{0(i)})(\hat{x}_{1(j)} - \Delta\hat{x}_{1(j)})(\hat{x}_{2(k)} - \Delta\hat{x}_{2(k)}) = 0.$$

Expanding this and ignoring second-order terms in $\Delta\hat{\boldsymbol{x}}_\kappa$, we obtain

$$\sum_{i,j,k,l,m=1}^{3} \epsilon_{ljp}\epsilon_{mkq}T_i^{lm}\Big(\Delta\hat{x}_{0(i)}\hat{x}_{1(j)}\hat{x}_{2(k)} + \hat{x}_{0(i)}\Delta\hat{x}_{1(j)}\hat{x}_{2(k)} + \hat{x}_{0(i)}\hat{x}_{1(j)}\Delta\hat{x}_{2(k)}\Big)$$
$$= \sum_{i,j,k,l,m=1}^{3} \epsilon_{ljp}\epsilon_{mkq}T_i^{lm}\hat{x}_{0(i)}\hat{x}_{1(j)}\hat{x}_{2(k)}. \qquad (*2)$$

From the definition of $\boldsymbol{x}_\kappa$ of Eq. (10.1), the third component of $\Delta\boldsymbol{x}_\kappa$ is identically 0. This constraint is written as

$$\sum_{k=1}^{3} k_i \Delta\hat{x}_{\kappa(i)} = 0, \qquad \kappa = 0, 1, 2, \qquad (*3)$$

where k_i is the ith component of $\boldsymbol{k} \equiv (0, 0, 1)^\top$. Dividing Eq. (∗1) by 2 and introducing Lagrange multipliers for Eqs. (∗2) and (∗3), we differentiate

$$\frac{1}{2}\sum_{\kappa=0}^{2} \|\tilde{\boldsymbol{x}}_\kappa + \Delta\hat{\boldsymbol{x}}_\kappa\|^2 - \sum_{i,j,k,l,m,p,q=1}^{3} \lambda_{pq}\epsilon_{ljp}\epsilon_{mkq}T_i^{lm}\Big(\Delta\hat{x}_{0(i)}\hat{x}_{1(j)}\hat{x}_{2(k)}$$
$$+ \hat{x}_{0(i)}\Delta\hat{x}_{1(j)}\hat{x}_{2(k)} + \hat{x}_{0(i)}\hat{x}_{1(j)}\Delta\hat{x}_{2(k)}\Big) - \sum_{\kappa=0}^{2}\sum_{i=1}^{3}\mu_\kappa k_i \Delta\hat{x}_{\kappa(i)}$$

with respect to $\Delta\hat{x}_{0(i)}$, $\Delta\hat{x}_{1(i)}$, and $\Delta\hat{x}_{2(i)}$. Letting the result be 0, we obtain

$$\Delta\hat{x}_{0(i)} = \sum_{j,k,l,m,p,q=1}^{3} \lambda_{pq}\epsilon_{ljp}\epsilon_{mkq}T_i^{lm}\hat{x}_{1(j)}\hat{x}_{2(k)} + \mu_0 k_i - \tilde{x}_{0(i)},$$

$$\Delta\hat{x}_{1(i)} = \sum_{j,k,l,m,p,q=1}^{3} \lambda_{pq}\epsilon_{ljp}\epsilon_{mkq}T_i^{lm}\hat{x}_{0(i)}\hat{x}_{2(k)} + \mu_1 k_i - \tilde{x}_{1(i)},$$

$$\Delta\hat{x}_{2(i)} = \sum_{j,k,l,m,p,q=1}^{3} \lambda_{pq}\epsilon_{ljp}\epsilon_{mkq}T_i^{lm}\hat{x}_{0(i)}\hat{x}_{1(j)} + \mu_2 k_i - \tilde{x}_{2(i)}.$$

Multiplying $\Delta\hat{\boldsymbol{x}}_\kappa$ by the projection matrix $\boldsymbol{P_k} = \boldsymbol{I} - \boldsymbol{k}\boldsymbol{k}^\top$ from left on both sides, and noting that $\boldsymbol{P_k}\Delta\hat{\boldsymbol{x}}_\kappa = \Delta\hat{\boldsymbol{x}}_\kappa$, $\boldsymbol{P_k}\tilde{\boldsymbol{x}}_\kappa = \tilde{\boldsymbol{x}}_\kappa$, and $\boldsymbol{P_k}\boldsymbol{k} = \boldsymbol{0}$, we obtain

$$\Delta\hat{x}_{0(s)} = \sum_{i,j,k,l,m,p,q=1}^{3} \lambda_{pq}\epsilon_{ljp}\epsilon_{mkq}T_i^{lm}P_{\boldsymbol{k}}^{si}\hat{x}_{1(j)}\hat{x}_{2(k)} - \tilde{x}_{0(s)} = \sum_{p,q=1}^{3} P_{pqs}\lambda_{pq} - \tilde{x}_{0(s)},$$

$$\Delta\hat{x}_{1(s)} = \sum_{i,j,k,l,m,p,q=1}^{3} \lambda_{pq}\epsilon_{ljp}\epsilon_{mkq}T_i^{lm}\hat{x}_{0(i)}P_{\boldsymbol{k}}^{sj}\hat{x}_{2(k)} - \tilde{x}_{1(s)} = \sum_{p,q=1}^{3} Q_{pqs}\lambda_{pq} - \tilde{x}_{1(s)},$$

$$\Delta\hat{x}_{2(s)} = \sum_{i,j,k,l,m,p,q=1}^{3} \lambda_{pq}\epsilon_{ljp}\epsilon_{mkq}T_i^{lm}\hat{x}_{0(i)}\hat{x}_{1(j)}P_{\boldsymbol{k}}^{sk} - \tilde{x}_{2(s)} = \sum_{p,q=1}^{3} R_{pqs}\lambda_{pq} - \tilde{x}_{2(s)},$$
$$(*4)$$

where P_{pqs}, Q_{pqs}, and R_{pqs} are defined by Eq. (10.11). Substituting these into Eq. (∗2), we obtain Eq. (10.13), using the definition of Eq. (10.12). Solving this for λ_{pq}, and substituting them into Eq. (∗4), we can determine $\Delta\hat{\boldsymbol{x}}_\kappa$. Hence, $\bar{\boldsymbol{x}}_\kappa$ is given by $\hat{\boldsymbol{x}}_\kappa - \Delta\hat{\boldsymbol{x}}_\kappa$. However, since second-order terms were ignored in Eq. (∗2), the trilinear constraint may not exactly hold. Hence, we newly regard this solution as $\hat{\boldsymbol{x}}_\kappa$ in the form of (10.14) and repeat the same procedure. The ignored terms decrease each time. In the end, $\Delta\hat{\boldsymbol{x}}_\kappa$ becomes $\boldsymbol{0}$, and the trilinear constraint exactly holds. From Eq. (∗1), the reprojection error E is given by Eq. (10.15). If its decrease is sufficiently small, we stop the iterations.

Bibliography

[1] G. H. Golub and C. F. Van Loan, *Matrix Computations*, The Johns Hopkins University Press, Baltimore, MD, 3rd ed., 1996, 4th ed., 2012. 37

[2] R. Hartley, In defense of the eight-point algorithm, *IEEE Transactions on Pattern Analysis and Machine Intelligence*, 19(6):580–593, 1997. DOI: 10.1109/34.601246. 82, 87

[3] R. Hartley and F. Kahl, Optimal algorithms in multiview geometry, *Proc. of the 8th Asian Conference on Computer Vision*, 1:13–34, Tokyo, Japan, 2007. DOI: 10.1007/978-3-540-76386-4_2. 99

[4] R. Hartley and P. Sturm, Triangulation, *Computer Vision and Image Understanding*, 68(2):146–157, 1997. DOI: 10.1006/cviu.1997.0547. 98

[5] R. Hartley and A. Zisserman, *Multiple View Geometry in Computer Vision*, 2nd ed., Cambridge University Press, Cambridge, UK, 2003. DOI: 10.1017/cbo9780511811685. 59, 82, 88, 97

[6] A. Heyden, R. Berthilsson, and G. Sparr, An iterative factorization method for projective structure and motion from image sequences, *Image and Vision Computing*, 17(13):981–991, 1999. DOI: 10.1016/s0262-8856(99)00002-5. 85

[7] K. Kanatani, *Group-Theoretical Methods in Image Understanding*, Springer, Berlin, 1990. DOI: 10.1007/978-3-642-61275-6. 32

[8] K. Kanatani, *Statistical Optimization for Geometric Computation: Theory and Practice*, Elsevier, Amsterdam, The Netherlands, 1996. Reprinted by Dover, New York, 2005. DOI: 10.1016/s0923-0459(96)x8019-4. 59, 98

[9] K. Kanatani, Latest progress of 3-D reconstruction from moving camera images, in X. P. Guo (Ed.), *Robotics Research Trends*, pages 33–75, Nova Science Publishers, Hauppauge, NY, 2008. 86

[10] K. Kanatani, Y. Sugaya, and H. Ackermann, Uncalibrated factorization using a variable symmetric affine camera, *IEICE Transactions on Information and Systems*, E89-D(10):2653–2660, 2006. DOI: 10.1093/ietisy/e90-d.5.851. 83

[11] K. Kanatani, Y. Sugaya, and K. Kanazawa, *Ellipse Fitting for Computer Vision: Implementation and Applications*, Morgan & Claypool, San Rafael, CA, 2016. DOI: 10.2200/s00713ed1v01y201603cov008. 60

[12] K. Kanatani, Y. Sugaya, and Y. Kanazawa, *Guide to 3D Vision Computation: Geometric Analysis and Implementation*, Springer, Cham, Switzerland, 2016. 59, 60, 73, 82, 84, 86, 88, 97, 99

[13] K. Kanatani, Y. Sugaya, and H. Niitsuma, Triangulation from two views revisited: Hartley-Sturm vs. optimal correction, *Proc. 19th British Machine Vision Conference*, pages 173–182, Leeds, UK, 2008. DOI: 10.5244/c.22.18. 98

[14] K. Kanatani, Y. Sugaya, and H. Niitsuma, Optimization without search: Constraint satisfaction by orthogonal projection with applications to multiview triangulation, *IEICE Transactions on Information and Systems*, E93-D(10):2386–2845, 2010. DOI: 10.1587/transinf.E93.D.2836. 98, 99

[15] Y. Kanazawa and K. Kanatani, Reliability of 3-D reconstruction by stereo vision, *IEICE Transactions on Information and Systems*, E75-D(10):1301–1306, 1995. 97, 98

[16] S. Mahamud and M. Hebert, Iterative projective reconstruction from multiple views, *Proc. IEEE Conference on Computer Vision and Pattern Recognition*, 2:430–437, Hilton Head Island, SC, 2000. DOI: 10.1109/cvpr.2000.854872. 85

[17] C. J. Poelman and T. Kanade, A paraperspective factorization method for shape and motion recovery, *IEEE Transactions on Pattern Analysis and Machine Intelligence*, 19(3):206–218, 1997. DOI: 10.1109/34.584098. 83

[18] W. H. Press, S. A. Teukolsky, W. T. Vetterling, and B. P. Flannery, *Numerical Recipes: The Art of Scientific Computing*, 3rd ed., Cambridge University Press, Cambridge, UK, 2007. 17, 41

[19] Y. Seo and A. Heyden, Auto-calibration by linear iteration using the DAC equation, *Image and Vision Computing*, 22(11):919–926, 2004. DOI: 10.1016/j.imavis.2004.05.004. 86

[20] H. Stewénius, F. Schaffalitzky, and D. Nistér, How hard is the three-view triangulation really? *Proc. of 10th International Conference on Computer Vision*, 1:686–693, Beijing, China, 2005. DOI: 10.1109/ICCV.2005.115. 98

[21] C. Tomasi and T. Kanade, Shape and motion from image streams under orthography—a factorization method, *International Journal of Computer Vision*, 9(2):137–154, 1992. DOI: 10.1007/bf00129684. 83

[22] B. Triggs, Autocalibration and the absolute quadric, *Proc. IEEE Conference on Computer Vision and Pattern Recognition*, pages 609–614, San Juan, Puerto Rico, 1997. DOI: 10.1109/cvpr.1997.609388. 86

Author's Biography

KENICHI KANATANI

Kenichi Kanatani received his B.E., M.S., and Ph.D. in applied mathematics from the University of Tokyo in 1972, 1974, and 1979, respectively. After serving as Professor of Computer Science at Gunma University, Gunma, Japan, and Okayama University, Okayama, Japan, he retired in 2013 and is now Professor Emeritus of Okayama University. He was a visiting researcher at the University of Maryland, U.S. (1985–1986, 1988–1989, 1992), the University of Copenhagen, Denmark (1988), the University of Oxford, U.K. (1991), INRIA at Rhone Alpes, France (1988), ETH, Switzerland (2013), University of Paris-Est, France (2014), Linköping University, Sweden (2015), and National Taiwan Normal University, Taiwan (2019). He is the author of K. Kanatani, *Group-Theoretical Methods in Image Understanding* (Springer, 1990), K. Kanatani, *Geometric Computation for Machine Vision* (Oxford University Press, 1993), K. Kanatani, *Statistical Optimization for Geometric Computation: Theory and Practice* (Elsevier, 1996; reprinted Dover, 2005), K. Kanatani, *Understanding Geometric Algebra: Hamilton, Grassmann, and Clifford for Computer Vision and Graphics* (AK Peters/CRC Press 2015), K. Kanatani, Y. Sugaya, and Y. Kanazawa, *Ellipse Fitting for Computer Vision: Implementation and Applications* (Morgan & Claypool, 2016), K. Kanatani, Y. Sugaya, and Y. Kanazawa, *Guide to 3D Vision Computation: Geometric Analysis and Implementation* (Springer, 2016), and K. Kanatani, *3D Rotations: Parameter Computation and Lie Algebra based Optimization* (AK Peters/CRC Press 2020). He received many awards including the best paper awards from IPSJ (1987), IEICE (2005), and PSIVT (2009). He is a Fellow of IEICE, IEEE, and IAPR.

Index